Book 3

Hormones in Reproduction

Reproduction in Mammals is designed as a basic
text to meet the needs of undergraduates reading
Biology, Zoology, Physiology, Medicine,
Agriculture and Veterinary Science, and as a
companion work to clinical textbooks of obstetrics
and gynaecology. It should also be of value to
more advanced students and to research workers
requiring a succinct treatment of topics in the
field of mammalian reproduction.

It is written in a lively style, copiously illustrated
with line drawings, and published as five slim
volumes. These deal with *Germ Cells and
Fertilization* and the reproductive cycles in
which these are involved: *Embryonic and Fetal
Development* from conception to parturition, and
its experimental manipulation: *Hormones in
Reproduction* and the way in which hormones
control the reproductive activities of the male and
female: *Reproductive Patterns* with a discussion of
species differences, reproductive behaviour, and the
effects of environmental factors: *Artificial
Control of Reproduction* with particular emphasis
on problems of human fertility and the ways in
which we may be able to combat them.

The aim throughout this series has been to
present an up-to-date and highly readable
comparative account of mammalian reproduction,
written by experts, in the belief that it is just as
important for the zoologist to understand something
about human reproduction as it is for the medical
student to have some fundamental knowledge of
reproductive biology.

REPRODUCTION IN MAMMALS

Book 3

Hormones in Reproduction

EDITED BY

C. R. AUSTIN

Fellow of Fitzwilliam College,
Charles Darwin Professor of Animal Embryology,
University of Cambridge

AND

R. V. SHORT

Fellow of Magdalene College,
Reader in Reproductive Biology,
University of Cambridge

ILLUSTRATIONS BY JOHN R. FULLER

CAMBRIDGE

AT THE UNIVERSITY PRESS 1972

Published by the Syndics of the Cambridge University Press
Bentley House, 200 Euston Road, London NW1 2DB
American Branch: 32 East 57th Street, New York, N.Y. 10022

Cambridge University Press 1972

Library of Congress Catalogue Card Number: 73–178279

ISBNS: 0 521 08438 5 hard covers
0 521 09696 0 paperback

Printed by offset in Great Britain by
Alden & Mowbray Ltd
at the Alden Press, Oxford

Contents

1696562

Contributors to Book 3

D. T. Baird,
Department of Obstetrics and Gynaecology,
39 Chalmers Street,
Edinburgh

B. A. Cross,
Department of Anatomy,
Medical School,
University Walk,
Bristol.

R. V. Short
Department of Veterinary Clinical Studies,
Madingley Road,
Cambridge.

R. B. Heap,
A. R. C. Institute of Animal Physiology,
Babraham,
Cambridge.

Alfred T. Cowie,
National Institute for Research in Dairying,
Shinfield,
Nr Reading.

Preface

Reproduction in Mammals is intended to meet the needs of undergraduates reading Zoology, Biology, Physiology, Medicine, Veterinary Science and Agriculture, and as a source of information for advanced students and research workers. It is published as a series of five small text books dealing with all major aspects of mammalian reproduction. Each of the component books is designed to cover independently fairly distinct sub-divisions of the subject, so that readers can select texts relevant to their particular interests and needs, if reluctant to purchase the whole work. The contents lists of all the books are set out on the next page.

This third volume in the series has to do with *Hormones in Reproduction* in man and other mammals. To begin with, attention is turned to the nature and properties of the various reproductive hormones, and then the present state of knowledge of the layout and function of the hypothalamus is reviewed. The remaining three chapters are on the endocrine control of sexual periodicity and breeding seasons, of pregnancy from implantation to parturition, and of mammary gland growth and the secretion of milk.

Books in this series

1 Reproductive hormones
D. T. Baird

Endocrinology is the study of endocrine glands and their secretory products, the hormones, which provide a chemical, as distinct from a nervous, transmission of information from one cell to another. As originally defined early in the century by Ernest Starling, hormones are secreted from specialized ductless or endocrine glands and transmitted via the blood stream to exert specific effects on cells in remote tissues. An example is follicle stimulating hormone (FSH) which is released from the anterior pituitary gland and stimulates the cells around the Graafian follicles in the ovary to grow and differentiate.

Other substances, e.g. histamine, produce effects on cells similar to those caused by classical hormones, but these only act locally on the surrounding tissue. Rapid destruction of these 'local hormones' prevents their circulation in the blood stream and provides one way of limiting the extent of their action.

Some hormones are formed locally within the target organ; their precursors or 'prehormones' are secreted by endocrine glands into the blood and taken up by the target organ. The prehormone itself may have very little intrinsic biological activity. For example, the ovaries and adrenal glands of women secrete the prehormone androstenedione, a biologically weak androgen which is formed in relatively large amounts; it is converted into the more active androgen testosterone by a variety of peripheral tissues. In this way testosterone can be produced in high concentrations locally in target organs without the necessity of high levels in circulating blood. Starling's original concept of hormones and endocrine glands has had to be extended to include these local hormones and prehormones. A hormone is better defined in terms of its action rather than its

TABLE 1-1

Endocrine Gland	Hormones	Main Target Organs
Pituitary		
Anterior lobe	Follicle Stimulating Hormone (FSH)	Ovary + testis
	Luteinizing Hormone (LH)	Ovary + testis
	Prolactin	Breast + ovary
Posterior lobe	Oxytocin	Breast + uterus
Ovary	Progesterone	Uterus + breast + brain
	Androstenedione	?muscle
	Oestradiol-17β	Uterus + breast + brain
Testis	Testosterone	Prostate, sebaceous glands, hair follicles etc. muscle, brain
Placenta	Chorionic gonadotrophin (CG)	?ovary
	Chorionic somatomammotrophin (CS)	?ovary

site of origin: it is a substance that induces growth, differentiation and/or alteration in the metabolic activity of cells.

The reproductive system is wholly dependent on hormones – in their absence, differentiation and development do not occur. Table 1-1 lists the endocrine glands and their hormones which influence reproductive function. The anterior lobe of the pituitary gland also secretes hormones that influence the function of other endocrine glands. This chapter will concentrate mainly on the classical reproductive hormones but some mention will also be made of prehormones and local hormones since they stir much current research interest.

TRANSPORT OF HORMONES IN THE BLOOD STREAM

Following production in their respective endocrine glands, protein hormones are often stored temporarily within the gland until their release is required. They are then secreted into the efferent venous capillaries draining the gland. Steroid hormones, on the other hand, are not stored in this way.

Many steroid hormones are transported in the blood stream strongly bound to plasma proteins. For example, oestradiol and testosterone are both bound to a globulin (sex-hormone-binding globulin), which is present in both male and female plasma. This provides a way of increasing the solubility of steroid hormones in aqueous media such as blood, as well as keeping the hormone in the blood and protecting it from metabolism by the liver. The steroid is probably biologically inactive when combined with the protein, but the steroid–protein complex is in equilibrium with free steroid, which is the form in which it is taken up by tissues (Fig. 1-1*a*). Because these binding proteins are highly specific for their respective steroid hormones, they have been used successfully for quantitative estimation of steroids in a system where one allows the hormone to be measured to compete with radioactively labelled steroid for the binding protein (Fig. 1-1*b*).

3

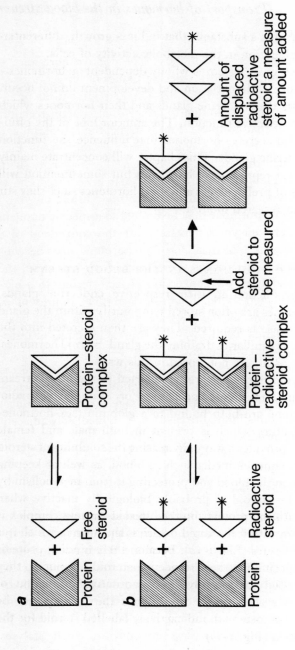

Fig. 1-1. *a* Reversible conjugation of a steroid to a binding protein in plasma
b Competition for the binding protein by radioactive and non-radioactive steroid can be used as an assay.

TARGET-ORGAN SPECIFICITY

One of the characteristics of most reproductive hormones is that they are concentrated in specific tissues or 'target organs'. The uptake of a hormone from the blood stream and its retention in a tissue may be limited to organs that contain receptor sites specific for that hormone. For example, the cytoplasm and nucleus of the cells in the endometrium of the uterus each contain a protein that binds oestradiol-17β very strongly (with high affinity). Thus although oestradiol has generalized effects throughout the body, it becomes concentrated in the uterus.

The action of a hormone may be limited to a specific tissue in other ways. The enzymes necessary for the conversion of a prehormone to the biologically potent hormone may be confined to certain target organs. For example, 5α-reductase, the enzyme that catalyses the conversion of testosterone to a biologically active metabolite, dihydrotestosterone, is present in high concentrations in the prostate and certain other male accessory sex glands. In this way the androgenic effect of testosterone is potentiated in specific tissues. The rapid destruction of *local* hormones ensures that the generalized systemic effect is reduced to a minimum: many prostaglandins are almost completely extracted from blood in one passage through the lungs, and so their action is limited to the tissues immediately adjacent to their site of production.

MODE OF ACTION OF HORMONES

The uptake of a hormone from the blood stream and its retention within the cells of the target organ is dependent on a receptor site specific for the particular hormone. Many hormones are known to increase the intracellular concentration of the nucleotide adenosine monophosphate ($3'$-$5'$-cyclic AMP) by influencing the adenyl cyclase enzyme; adrenocorticotrophic hormone (ACTH), vasopressin, luteinizing hormone (LH), adrenaline and glucagon all work in this way. This enzyme, which catalyses

5

the conversion of adenosine triphosphate (ATP) to 3′-5′-cyclic AMP, is situated on the inner surface of the cell membrane and may be identical with what we have referred to as the specific receptor site. 3′-5′-cyclic AMP is thought to act as a 'second messenger', by propagating the effect of the hormone ('first messenger') throughout the cell (Fig. 1-2). Hormone

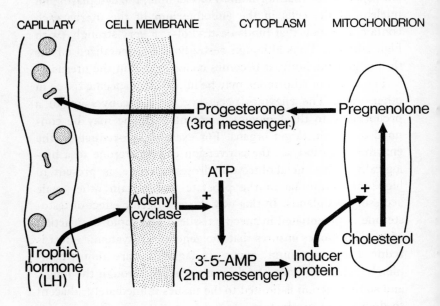

Fig. 1-2. The way in which a trophic hormone acts on the cell to stimulate the production of cyclic AMP, which then acts as a 'second messenger' to stimulate steroid secretion.

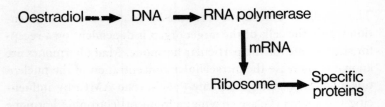

Fig. 1-3. The way in which oestradiol-17β stimulates protein synthesis.

6

action may include the production of a 'third messenger' – e.g. LH may increase production by the ovary of progesterone which then influences the function of other cells.

Almost all known hormones have been found to influence protein synthesis. Hormones may activate genes which induce the production of new types of messenger RNA which in turn promote the synthesis of specific proteins (Fig. 1-3).

PITUITARY GONADOTROPHINS

The anterior pituitary gland secretes three protein hormones that have a critical influence on the reproductive system. These gonadotrophic hormones are FSH, LH, and prolactin. Other hormones of the anterior pituitary, e.g. ACTH and thyroid-stimulating hormone (TSH), also influence the reproductive system indirectly but cannot be considered as true reproductive hormones. None of these hormones has been isolated in chemically pure form, but considerable progress has been made in identifying separate amino-acids and their sequence. Three biologically different gonadotrophic fractions have been isolated from the pituitary glands of many mammals.

Structure

FSH and LH are glycoproteins with molecular weights around 30 000. The carbohydrate content (27 per cent for FSH) and, in particular, the sialic acid content (FSH 5 per cent and LH 1.4 per cent) is essential for their biological action. Both FSH and LH are composed of two sub-units, one of which is common to both gonadotrophins and to TSH. This common sub-unit explains the degree of immunological cross-reaction that occurs between these three pituitary hormones. Prolactin is probably a single polypeptide chain with about 211 amino-acid residues and a molecular weight about 24 000.

Site of production

Gonadotrophins are produced in the anterior lobe of the pituitary gland under control of 'releasing factors' transported from the hypothalamus via the hypothalamic–hypophyseal portal system (see Chapter 2). FSH and LH are secreted by the *basophilic* staining cells of the pituitary, while the *acidophilic* cells produce prolactin (and growth hormone). The hormone is concentrated in granules 200–350 μm in diameter which can be identified on light microscopy in the cytoplasm of the appropriate cell. These granules probably represent a store of preformed hormone because they disappear when a large amount of hormone is being released from the gland.

Biological action

The biological action of these hormones will be discussed in more detail in Chapter 3. It is important to appreciate that in physiological conditions no single hormone is likely to be secreted exclusively and that the ultimate biological activity is determined by the relative proportions of a number of hormones, together with the morphological state of the target organ.

In the female, FSH will promote the growth and development of follicles in the ovary; it is thought to stimulate the development of follicles from the antral to the pre-ovulatory stage, and will produce certain biochemical changes, such as increased oxygen uptake and protein synthesis, especially in the theca cells. In biologically pure form FSH alone will not promote steroid secretion from the developing follicles. In the male, FSH stimulates growth of the seminiferous tubules as well as being important in the first stages of spermatogenesis.

LH has several distinct actions in the female. First, it stimulates steroid synthesis by all cell types in the ovary (corpus luteum, theca, granulosa and interstitium), and a certain basal concentration of LH is probably a necessary condition of steroid synthesis in the ovary. Its main action is stimulating the conversion of cholesterol to pregnenolone in a manner very similar to the stimulation of adrenal steroid synthesis by ACTH (see Fig. 1-2). Following the administration of LH the ovary is depleted of cholesterol (and ascorbic acid) which is mobilized for increased steroid synthesis. LH also induces a rapid and large increase in ovarian blood flow. This hyperaemic effect corresponds to the increase in blood flow produced by other trophic hormones on their specific target organs, e.g. ACTH on the adrenal, TSH on the thyroid, and secretin on the pancreas. By providing an increased supply of certain critical metabolites, hyperaemia is probably an integral part of the mode of action of trophic hormones on their target organs.

In addition, LH will induce ovulation in follicles suitably primed with FSH. The ovulatory action of LH is separate from its steroidogenic action and is controlled quite differently. The surge of pituitary LH release responsible for ovulation is produced by the positive feedback of oestradiol on the hypothalamus, causing a discharge of LH releasing factor (see Chapter 2).

In the male, LH stimulates increased synthesis and secretion of testosterone from the interstitial or Leydig cells in the testis. The androgen so produced is necessary for the development of male secondary sexual characteristics and for the final maturation of spermatozoa.

Prolactin has two distinct actions in most mammals. First, together with ACTH and growth hormone (GH) it is part of the pituitary 'lactogenic complex'. It acts synergistically with oestrogen on the duct system of the mammary gland and with progesterone on the lobular–alveolar system. In addition, together with corticosteroids it initiates and maintains milk secretion in the developed breast (see Chapter 5). Large increases in the concentration of prolactin in peripheral blood occur during milking in cows and sheep. Paradoxically, prolactin is also released by the male during intercourse, although it has no known function in males. Secondly, prolactin, together with LH, is part of the pituitary 'luteotrophic complex', being responsible for maintaining the structure and function of the corpus luteum in many mammals. This luteotrophic function has been demonstrated in sheep and rats, and its function in

9

Reproductive hormones

women will be better recognized now that a specific radio-immunoassay has been developed for human prolactin.

POSTERIOR PITUITARY HORMONES

The posterior pituitary gland secretes two octapeptide hormones, vasopressin or antidiuretic hormone, and oxytocin, with molecular weights of approximately 1100. The structures of both vasopressin and oxytocin have been elucidated and the hormones synthesized (Fig. 1-4). Their structures are very

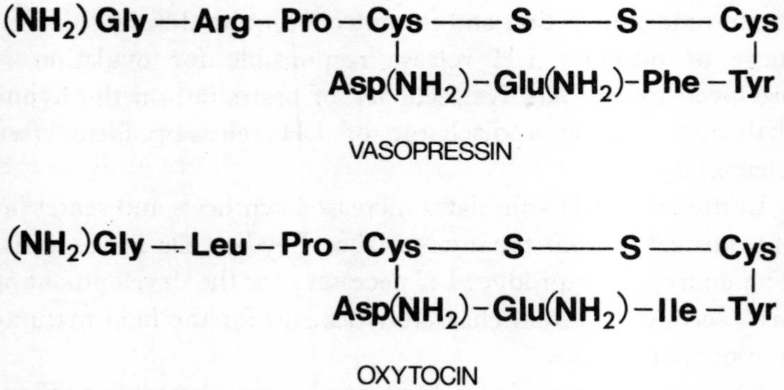

$$(NH_2)\,Gly - Arg - Pro - Cys \underline{\hspace{1cm}} S \underline{\hspace{1cm}} S \underline{\hspace{1cm}} Cys$$
$$Asp(NH_2) - Glu(NH_2) - Phe - Tyr$$

VASOPRESSIN

$$(NH_2)Gly - Leu - Pro - Cys \underline{\hspace{1cm}} S \underline{\hspace{1cm}} S \underline{\hspace{1cm}} Cys$$
$$Asp(NH_2) - Glu(NH_2) - Ile - Tyr$$

OXYTOCIN

Fig. 1-4. The two peptide hormones produced by the posterior pituitary.

similar and their properties overlap considerably. Some non-mammalian vertebrates produce only one hormone from the posterior pituitary gland, vasotocin, which has actions intermediate between oxytocin and vasopressin.

Oxytocin is synthesized in the paraventricular nucleus of the hypothalamus and transported along the axons of the supra-optico–hypophyseal nerve tract to the neighbourhood of blood spaces in the posterior pituitary. Granules containing oxytocin combined with a specific binding protein, 'neurophysin', are stored in the posterior pituitary. Oxytocin has two distinct actions. First, it causes contraction of the smooth muscle of the uterus. It is rapidly destroyed in blood by an enzyme (oxytocinase) but is present in increased amounts in blood during labour. This property of oxytocin is widely used in obstetrics to induce labour and in the post-partum period to control uterine haemorrhage. Distension of the vagina during parturition can produce a reflex release of oxytocin. Secondly, oxytocin stimulates the myoepithelial cells in the mammary gland and so causes milk ejection. Its release from the pituitary is stimulated by suckling, the afferent pathway being via the sensory nerves originating in the nipple.

The factors stimulating the release of oxytocin from the posterior pituitary also cause vasopressin to be secreted. When administered in large doses, oxytocin also has a weak antidiuretic action so that water retention occurs. The copious flow of urine that follows the consumption of beer is well known, and alcohol is an effective inhibitor of vasopressin. Alcohol will also inhibit oxytocin release, and this fact is made use of clinically in the treatment of threatened premature labour.

STEROID HORMONES

The ovary, testis and adrenal glands secrete a class of lipid compounds known as steroids which have a common basic structure. The steroid nucleus is composed of a cyclopentanophenanthrene ring such as in cholesterol (Fig. 1-5), the separate hormones differing in the nature of the attached side chains. All the steroid-secreting endocrine glands make their hormones from a basic molecule (acetate) containing only two carbon atoms. The biosynthetic steps via cholesterol to pregnenolone are common for all steroid hormones including the corticosteroids.

The enzymes responsible for catalysing the chemical steps involved in the biosynthesis of sex hormones are present in adrenal, testis and ovary. The pattern of steroids secreted by the respective endocrine glands is determined by the relative proportions of cell types, the anatomical organization of the gland, the blood supply, the concentration of co-factors and precursors present in the gland, and the presence of trophic stimuli.

Fig. 1-6 represents the main routes of biosynthesis of steroid hormones. Progesterone is formed by oxidation of pregnenolone. It is the hormone responsible for maintenance of pregnancy in most mammals. It causes relaxation of smooth muscle throughout the body, including the gall bladder and the gastrointestinal tract, and reduces the excitability of the myometrium. Together with oestrogen it causes extensive growth and development of the lobular–alveolar system of the breast and of the

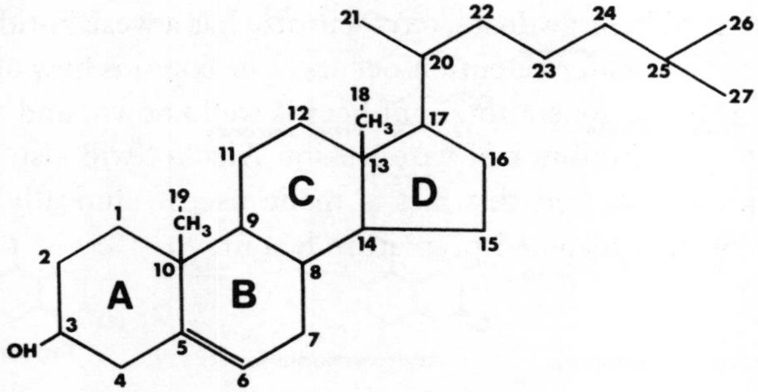

Fig. 1-5. The formula of cholesterol, showing the convention for numbering the rings and the carbon atoms.

endometrium of the uterus. Progesterone is secreted in large amounts by the corpus luteum, and in smaller quantities by the granulosa cells of the ovarian follicle before ovulation and by the adrenal.

Pregnenolone and progesterone are converted to androgens such as dehydroepiandrosterone, androstenedione and testosterone, by hydroxylation at position 17 followed by removal of the C_{21} side chain (Fig 1-7). Androgens are responsible for the development of male characteristics. In fetal life the testis produces an androgenic substance which promotes the development of the male genital organs. In adult life, androgen secretion from the testis controls the development and maintenance of the male secondary sex characteristics – deepening of voice, sexual hair, enlargement of the genitalia and accessory glands,

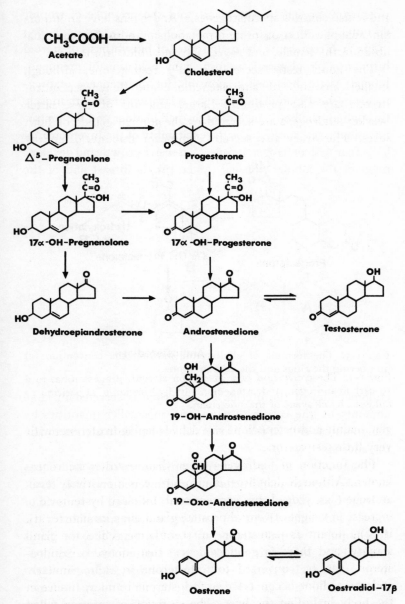

Fig. 1-6. The main routes of biosynthesis of androgens, oestrogens and progesterone from acetate and cholesterol.

and it also controls spermatogenesis. Androgens have an important anabolic effect on protein metabolism and influence sexual libido in the female, and aggression and libido in the male.

The adult testis secretes mainly testosterone, although smaller amounts of androstenedione and dehydroepiandrosterone are also produced. Larger amounts of these latter 'weaker' androgens are secreted by the adrenal glands of both sexes. The ovary also secretes significant amounts of andro-

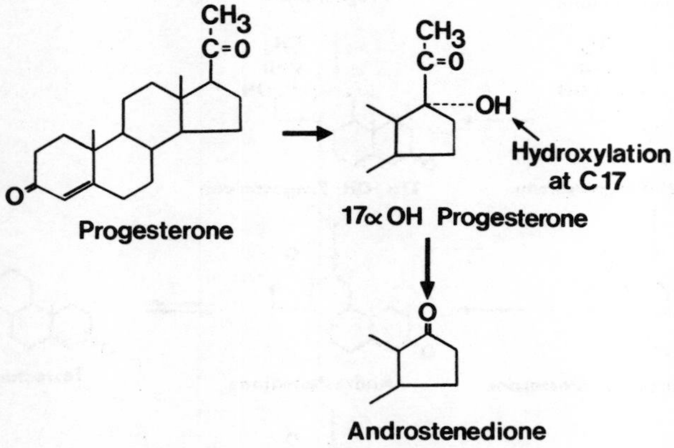

Fig. 1-7. The conversion of a 21-carbon steroid, progesterone, to a 19-carbon androgen, androstenedione, by hydroxylation at position 17 followed by cleavage of the side chain.

gen, mainly androstenedione and dehydroepiandrosterone, with very little testosterone.

The function of androgens other than testosterone is unknown. Although dehydroepiandrosterone is a relatively weak androgen as judged by conventional bioassay systems (e.g. increase in weight of ventral prostate gland of an immature rat), it is as potent as testosterone in stimulating sebaceous gland activity, and there is good evidence that dehydroepiandrosterone can be converted to testosterone in skin. Similarly androstenedione is converted to testosterone in many tissues in the body including the liver. The secretion of these so-called 'weak androgens' into the blood stream and their conversion to testosterone in target organs may be a means of providing high concentrations of androgen at specific tissues without a high concentration of testosterone in blood (Fig. 1-8). In this way inappropriate side effects, such as androgenization in women, may be avoided.

More recently, evidence has accumalated that testosterone itself is converted into a potent metabolite (dihydrotestosterone)

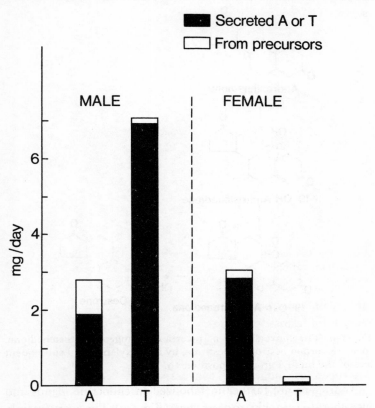

Fig. 1-8. Daily production of androstenedione (A) and testosterone (T) in men and women. Some is secreted directly by the adrenal, ovary or testis, and some is produced in the circulation from precursors secreted by the endocrine glands. (From D. Baird, R. Horton, C. Longcope, and J. F. Tait, *Perspectives in Biology and Medicine* **11,** 384, Fig. 2, University of Chicago Press. Copyright 1968 by the University of Chicago.)

in the nuclei of cells within the target organ (e.g. prostate). Thus even in the male where high levels of testosterone in peripheral blood are desirable, the biological effect of the hormone is potentiated further in the target organ. If dihydrotestosterone is regarded as the active hormone, dehydroepiandrosterone, androstenedione and testosterone can all be considered as prehormones.

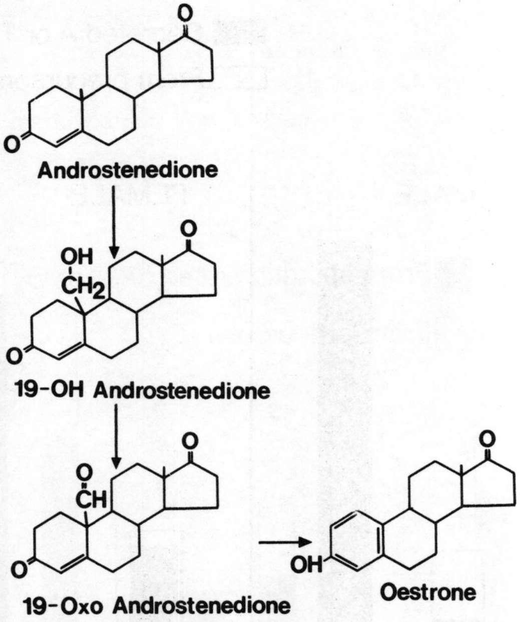

Fig. 1-9. The conversion of a 19-carbon androgen, androstenedione, to an 18-carbon oestrogen, oestrone, by hydroxylation and subsequent loss of the methyl group at position 19.

Androgens are further metabolized in endocrine glands and elsewhere (e.g. in skin) to oestrogens (Fig. 1-9). The ovary (mainly theca cells of the follicle, and in women the corpus luteum) secretes large amounts of oestradiol-17β. Oestrogens are responsible for the development of the secondary sex characteristics of the female, including growth of the duct system of the breast, the uterus and vaginal epithelium. In most species, oestradiol-17β, which is the most potent naturally occurring oestrogen, is the main oestrogen secreted by the ovary. It is concentrated in specific target organs (e.g. the uterus) where it is bound strongly to a protein occurring in the cytoplasm of the cells. The oestradiol is then transported through the cytoplasm to the nucleus, where it is bound to a second protein and influences the metabolic functions of the cell including protein synthesis. Throughout this process the oestradiol-17β is protected from further metabolism.

Oestrogens are secreted in small amounts by the adrenal and testis as well as by the ovary. However, in women after the

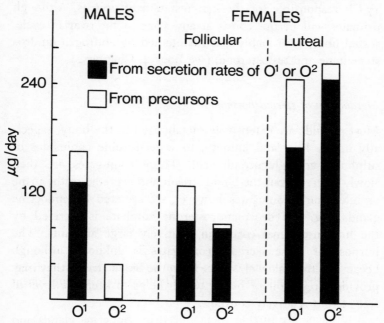

Fig. 1-10. Daily production of oestrone (O[1]) and oestradiol-17β (O[2]) in men, and in women during the follicular and luteal phases of the menstrual cycle. Some oestrogens are secreted directly by the adrenal, ovary or testis, and some are produced in the circulation from precursors secreted by the endocrine glands. (From D. Baird, R. Horton, C. Longcope, and J. F. Tait, *Perspectives in Biology and Medicine* 11, 384, Fig 3, University of Chicago Press. Copyright 1968 by the University of Chicago.)

Reproductive hormones

menopause, and in men, a greater amount of circulating oestrogen is derived not by direct secretion but by metabolism of androgens, e.g. androstenedione and testosterone, in tissues other than endocrine glands (Fig. 1-10). In this way although the adrenal secretes insignificant amounts of testosterone and oestradiol-17β it is responsible for oestrogenic as well as androgenic biological effects.

Steroid hormones have significant effects on behaviour. Thus, testosterone secretion from the testis maintains sexual desire in the male, and initiates sexual libido at puberty and at the onset of the breeding season in some species. Oestradiol-17β is responsible for the manifestation of oestrus. Although primates will permit coitus at any stage of the ovarian cycle, sexual libido is probably also maintained by androgen (androstenedione and testosterone) (see Book 4, Chapter 2).

Metabolism of steroid hormones

Most steroids are extensively metabolized in the body, especially in the liver and kidneys, to water-soluble conjugates of sulphuric and glucuronic acid. These conjugates are then slowly cleared from the blood stream and excreted in the urine or bile. Some conjugates however are secreted by endocrine glands, e.g. dehydroepiandrosterone sulphate is secreted by the human adrenal cortex in relatively large amounts. The purpose of these secreted conjugates is unknown although because of their slow clearance from the blood stream they may provide a reservoir or buffer which helps to keep the level of free steroid relatively constant.

Cellular origins of steroid hormones

Most testicular steroid hormones are produced by the Leydig or interstitial cells of the testis. The epididymal cells can metabolize steroids secreted by the Leydig cells to other active hormones. The Sertoli cells which are present within the seminiferous

18

tubules may secrete androgens and other steroids into the lumen of the tubule. Although the number of these cells is probably a fraction of the Leydig cells, the steroids so produced may have significant local effects on seminiferous tubules and hence on spermatogenesis.

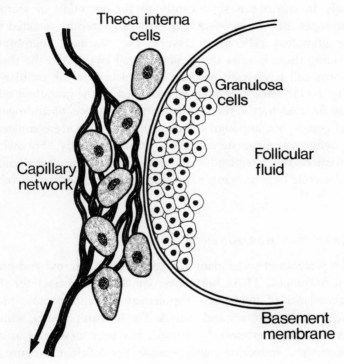

Fig. 1-11. The structure of the wall of the Graafian follicle, showing how the granulosa cells are deprived of a blood supply by the basement membrane.

All cell types in the ovary have the capacity to make steroid hormones. Oestrogens are largely secreted by the theca cells of the follicle, and in primates by the theca lutein cells of the corpus luteum. Progesterone is produced by the granulosa cells of the follicle and more especially by the luteinized granulosa cells of the corpus luteum. Androgens are probably secreted by the theca cells of the follicle and corpus luteum, as well as by the interstitial and hilar cells.

In the ovary the steroid hormones secreted by a particular cell type are determined by the stage of the ovarian cycle. Prior to ovulation the granulosa cells within the follicle cavity are unlikely to contribute significantly to the secretion of steroid hormones into the ovarian vein. Any progesterone secreted by the granulosa cells must traverse the basement membrane dividing theca interna and granulosa cell layers, and the theca interna cell layer itself before gaining access to the capillaries (Fig. 1-11). Some progesterone produced by the granulosa cells may be used by the theca cells for the synthesis of androgens and oestrogens, although the theca interna cells are competent to synthesize progesterone from acetate directly. Following ovulation, the luteinized granulosa cells become vascularized and secrete large amounts of progesterone into the ovarian vein.

PLACENTAL HORMONES

The placenta of some mammals secretes both steroid and protein hormones. These hormones, which are necessary for the maintenance of pregnancy, supplement or replace those produced by the pituitary and ovaries. The human placenta, which has been most extensively studied, has been described as an 'incomplete' endocrine gland because it is deficient in many of the essential enzymes necessary for complete synthesis of steroid hormones from acetate. Recognition of the importance of the human fetus in the production of placental hormones has led to the use of the term 'feto-placental unit'. More accurately, it should be called the feto-placento-maternal unit, since both the maternal and fetal liver and adrenal glands supply preformed precursors to the placenta, which transforms them into more active hormones. Whether this complex interrelationship applies to other mammalian species remains to be determined; it does not seem to be true of the Rhesus monkey, for example.

Steroids

The human placenta is unable to synthesize significant quantities
of steroid hormones from acetate. It probably converts cholesterol or pregnenolone from the maternal blood to progesterone
which is secreted by the placenta in large amounts and is important in maintaining pregnancy in many species. In addition to
inducing widespread adaptations in maternal physiology, pro-

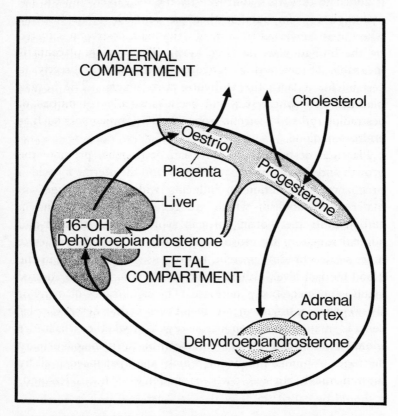

Fig. 1-12. The human feto-placental unit, showing how the mother
provides precursor (cholesterol) to the placenta, which converts it to
progesterone for release into the maternal and fetal circulations. In
the fetus, it is further metabolized by the adrenal cortex, the liver and
the placenta to oestriol, which is passed back into the maternal circulation for excretion in the urine.

Reproductive hormones

gesterone reduces the excitability and contractility of uterine muscle. In this way it probably helps to prevent premature expulsion of the fetus from the uterus.

Progesterone is also secreted from the placenta into the fetus via the umbilical vein. It is converted by the fetal adrenal to adrenocortical steroids, as well as a variety of C_{19} androgens such as dehydroepiandrosterone. An additional hydroxyl group is added to this weak androgen by the fetal liver, and the 16-hydroxylated compound is then passed back to the placenta where it is converted to oestriol – the main oestrogen secreted by the human placenta (Fig. 1-12). The human placenta is incapable of converting dehydroepiandrosterone directly to oestriol for it lacks the 16-hydroxylase enzyme. The human placenta also synthesizes and secretes smaller quantities of oestradiol-17β and oestrone from other C_{19} precursors such as androstenedione.

Placental oestrogens, like ovarian oestrogens, promote the growth and development of the breast and the uterine muscle in pregnancy, in addition to inducing widespread changes in maternal metabolism which are reflected by alterations in carbohydrate metabolism, protein synthesis, and thyroid and adrenal function. By causing the maternal liver to synthesize more steroid-binding protein, oestrogens produce a rise in the blood cortisol level, although cortisol secretion by the adrenal gland is not necessarily increased. The significance of many of these changes is unknown; we do not even know whether the placenta is an autonomous endocrine organ, or whether it is under some sort of trophic control. And oestriol, the principal urinary oestrogen of human pregnancy, which is not produced at all by most mammals, is a steroid whose function we do not understand.

Protein hormones

The human placenta also secretes large quantities of at least two protein hormones, human chorionic gonadotrophin (HCG) and human chorionic somatomammotrophin (HCS). There is a possibility that the placenta may also produce a protein hormone with thyrotrophic properties, although it has not been positively identified. The placenta of other mammals such as the rat and sheep also produce protein hormones which are luteotrophic since they maintain the life of the corpus luteum. The serum of pregnant mares contains very high concentrations of a gonado-trophic substance (pregnant mare's serum gonadotrophin or PMSG) which is secreted by the endometrial cups of the pregnant uterus (see Book 4, Chapter 1).

Chorionic gonadotrophin. The urine and serum of pregnant women contain high concentrations of HCG. Its detection in human urine was the basis of the earliest biological tests for the diagnosis of pregnancy; extracts of urine from pregnant women injected into immature mice, rabbits, rats or toads induced stimulation of the ovaries, ovulation, and follicular haemorrhages. Nowadays HCG is measured by a rapid and simple immunological assay. Chimpanzees and rhesus monkeys also produce a chorionic gonadotrophin.

HCG is a glycoprotein of molecular weight around 30 000. Its chemical structure and properties are very similar to those of pituitary LH although it has a higher carbohydrate content (8.5 per cent sialic acid). Because of this, it is relatively resistant to metabolic degradation and hence has a much longer biological half life than LH. It is composed of at least two sub-units, one of which is common to FSH, LH and TSH, accounting for the immunological similarity of these protein hormones. HCG is immunologically indistinguishable from LH with the antisera currently available.

HCG is produced by the trophoblast (both syncytio- and cyto-trophoblastic cell layers). In addition to normal placentae, it is also produced by hydatidiform moles, certain placental tumours such as choriocarcinoma, and rare ovarian and testicular tumours containing trophoblastic tissue.

Although HCG will reproduce many of the actions of LH when injected into experimental animals or human subjects, its

23

physiological function is still not defined. It will stimulate steroid synthesis when added to ovarian or testicular tissue *in vitro*, as well as promoting increased secretion of steroid hormones by the gonads *in vivo*. It will stimulate the conversion of androgens to oestrogens (aromatization) by the human placenta. Whether the latter action of HCG is important in regulating the placental production of oestrogens is unknown. HCG will also induce ovulation in follicles suitably primed with FSH. HCG together with HCS is probably part of the luteotrophic complex which converts the corpus luteum of the cycle into one of pregnancy.

Chorionic somatomammotrophin. Extracts of human placenta contain a fraction with many of the properties of both growth hormone and prolactin. The original name, human placental lactogen (HPL), has now been replaced by the more accurate though rather ponderous term, human chorionic somatomammotrophin (HCS). It is synthesized by the syncytiotrophoblast cells of the placental villi, and is secreted almost exclusively into the maternal circulation in large amounts (about 1g per day). It is similar chemically and immunologically to human growth hormone, having a molecular weight of around 20 000. Although the half-life is quite short, the secretion rate is so great that the concentration of HPL rises 1000-fold throughout pregnancy. The placentae of the monkey and rat also secrete chorionic somatomammotrophins that are immunologically distinct from HCS.

HCS promotes development of the breast, and together with HCG is probably responsible for the maintenance of the corpus luteum of pregnancy. It produces changes in protein and carbohydrate metabolism similar to those of growth hormone. Neither the function of these changes nor the factors controlling the production of HCS are known.

Pregnant mare's serum gonadotrophin (PMSG). The endometrial 'cups' of the pregnant uterus of the horse and donkey secrete large amounts of the protein hormone PMSG, the molecular weight of which is larger than that of any other gonadotrophin (about 70 000). The high molecular weight together with the very high carbohydrate content (49 per cent with 10.4 per cent sialic acid) impairs its clearance from the circulation and accounts for its long half life and high concentration. It has both FSH- and LH-like properties, but the former predominates. Its physiological function is unknown, and its secretion is regulated by the genotype of the fetus (see Book 4, Chapter 1).

PROSTAGLANDINS

Prostaglandins are a group of biologically active lipids which appear to play important roles in many reproductive processes. In 1930 two New York gynaecologists noted that the human uterus would react with strong contractions or relaxation in response to fresh human semen. Von Euler in Sweden found that lipid extracts of the semen of other species (monkey, sheep and goat) and extracts of the seminal vesicles of sheep stimulated strong contractions in smooth muscle. He named the lipid-soluble acid fraction which contained this biological activity 'prostaglandin'. Since then prostaglandins (PG) have been extracted from almost every tissue in the body, and have been found to possess a variety of chemical structures and biological activities. Three groups of prostaglandins are recognized, A, E and F. All are lipid-soluble, unsaturated hydroxyacids containing 20 carbon atoms (Fig. 1-13). The basic 'prostanoic acid' skeleton, common to all the prostaglandins, is formed from naturally occurring essential fatty acids. Bishomo-γ-linolenic acid and arachidonic acid are converted to PGE_1 and PGE_2 by extracts of lung, brain, endometrium, uterus and stomach, as well as by other tissues. The widespread distribution of the enzymes responsible for the biosynthesis of prostaglandins is reflected by the fact that they are found (in low concentrations: 1 μg/g wet tissue) in lung, thymus, brain, kidney, umbilical cord, uterus, amniotic and menstrual fluid. The highest concentrations occur in human and sheep seminal plasma (about 100 μg/ml). The enzymes responsible for the biosynthesis of prostaglandins from essential fatty acids are present in the microsomal fraction of the cell and require molecular oxygen and a reducing agent. The three oxygen

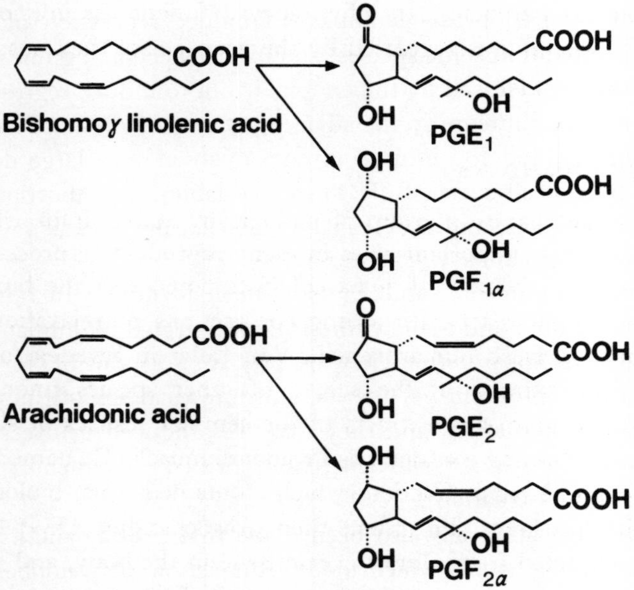

Fig. 1-13. Structure and biosynthesis of some of the prostaglandins.

atoms introduced into the molecular structure at carbon atoms 9, 11 and 15 are derived from molecular oxygen.

The actions of the prostaglandins of the A, E and F series are often dissimilar and opposing. They have profound effects on smooth muscle, including that of the intestine and the uterus. PGEs usually depress the motility of the Fallopian tube and uterus, whereas PGFs have the opposite action. Because they are rapidly removed from the circulation, they probably produce effects mainly within or immediately adjacent to the tissue where they are produced; in other words, they are local hormones. The importance of the very high concentration of prostaglandins in human semen is unknown, although they may be concerned in sperm transport. They are found in relatively large amounts in the maternal part of the human placenta and in amniotic fluid from women in labour. They may play an important part in labour by virtue of their oxytocic properties (see Book 2, Chapter 3).

Administered prostaglandins will induce regression of the corpus luteum in some species, and in the sheep and guinea pig they may be released by the endometrium to induce regression of the corpus luteum in the adjacent ovary (luteolysis). Prostaglandins E_1 and $F_{2\alpha}$ infused intravenously in very large doses have been used successfully to induce labour and abortion in human subjects, and they can be given locally, e.g. as vaginal suppositories (Book 5, Chapter 2).

Prostaglandins $F_{2\alpha}$ and E_2 have been identified in human menstrual fluid and their synthesis demonstrated by human endometrial strips *in vitro*. They are probably responsible for the increased contractility of uterine and intestinal smooth muscle during menstruation, and they may be produced in excessive amounts by women suffering from primary dysmenorrhoea (painful menstruation).

The suggestion has also been made that prostaglandins act as controlling agents which modify the action of hormones on target cells. The effects of many hormones that increase the activity of the adenyl cyclase system are antagonized or enhanced *in vitro* and *in vivo* by PGE_1 or PGE_2. LH, for example, by mobilizing phospholipids and triglycerides in the ovary, might increase the synthesis of prostaglandins through a greater concentration of their precursor unsaturated fatty acids. The prostaglandins so synthesized could modify the metabolic consequences of LH on the ovarian cell.

This chapter has attempted to summarize the chemical nature and properties of hormones that affect reproductive function. The classical hormones released into the blood stream initiate

Reproductive hormones

a series of events at the target organ involving the release of further chemical messengers or local hormones. Endocrinology is concerned with the mechanism of action of hormones at a cellular level as well as with the control of their biosynthesis and secretion into the blood stream. Although the relationship between the endocrine glands and the reproductive system has been recognized for over fifty years, there are still considerable gaps in our understanding of the intricate mechanisms.

SUGGESTED FURTHER READING

Foetus and Placenta. Ed. A. Klopper and E. Diczfalusy. Oxford and Edinburgh; Blackwell Scientific Publications (1969).
Metabolism of Steroid Hormones. Ed. R. I. Dorfman and F. Ungar. New York and London; Academic Press (1965).
The Chemistry of the Gonadotrophins. W. R. Butt. Springfield; Thomas (1967).
Hormones, Cells and Organisms. P. C. Clegg and A. G. Clegg, London; Heinemann (1969).
Hormone Chemistry. W. R. Butt. London; Van Nostrand (1967).
The prostaglandins; a family of biologically active lipids. S. Bergström, L. A. Carlson and J. R. Weeks. *Pharmacological Reviews* **20,** 1–48 (1968).

2 The hypothalamus
B. A. Cross

Doing what comes naturally always involves the central nervous system but with reproduction this is more true than many realize. For the human male it is self-evident that libido and sexual arousal involve a conscious element, while erection, seminal emission and ejaculation have the character of reflex nervous events albeit under some cerebral control. In the female, fluctuations of sexual feeling within the menstrual cycle also have a nervous origin. Less obvious is the fact that the menstrual cycle itself, including the recurring sequence of ovarian changes, ovulation and the secretion of gonadotrophic hormones is subordinated to the brain. For many years after the discovery of the gonadotrophic hormones FSH and LH, the anterior pituitary gland, as the 'conductor of the endocrine orchestra', was considered an autonomous organ, and the female cycle was thought to result from the interlocking secretory rhythms of gonadotrophic and gonadal hormones. This idea gradually succumbed to the weight of evidence from numerous sources. Precocious puberty in human patients, for example, may be associated with lesions in the brain rather than directly in the pituitary. F. H. A. Marshall of Cambridge accumulated evidence from observations in animals that breeding seasons were influenced by external stimuli which could only exert their effect on the reproductive organs through the mediation of the nervous system; even ovulation in some animals (e.g. rabbit, cat, ferret and mink) necessitates the nervous stimulus of coitus (Book 1, Chapter 4). Further, many controlled experimental studies in animals show that the placement of localized lesions in the hypothalamus or electrical stimulation of that structure can dramatically affect reproductive processes.

The hypothalamus

The hypothalamus is a very inconspicuous part of the human brain, visible on the undersurface merely as an oval band of grey matter suspending the pituitary gland. Though its deeper portions are more extensive the total weight is only about 5g, i.e. one three-hundredth part of the whole brain. Yet there is good reason to believe that this small region controls not only sexual cycles but also other pituitary functions in growth, pregnancy, lactation and stress, as well as temperature regulation, water balance, sleep and various emotional reactions.

In the medial portions of the hypothalamus, which form the walls of the third ventricle of the brain, the nerve cells are aggregated for the most part into a number of bilaterally paired groups, known as nuclei. Among the most readily distinguished of these are the mammillary, dorsomedial, ventromedial, paraventricular, supraoptic, suprachiasmatic and arcuate nuclei (see Fig. 2-1). Other cell groupings, less clearly demarcated, are the posterior hypothalamic area, anterior hypothalamic area and preoptic area (Fig. 2-1). Knitting together these various constellations of cells is a dense matrix of fine nerve fibres, the details of whose ramifications are largely unknown. It is certain, however, that many fibres gain access to the highly cellular medial zone from the lateral hypothalamic areas, which though more sparsely populated with cell bodies convey large numbers of ascending and descending nerve fibres, including some prominent bundles of thick nerve fibres (myelinated axons).

So far, this anatomical description would seem to hold little prospect of stirring physiological interest. The clue to its remarkable diversity of action seems to lie in the nature of the functional connections of the hypothalamus. Unlike any other brain region it not only receives sensory inputs from almost every other part of the central nervous system including the cerebral cortex, basal ganglia, thalamus, midbrain and hindbrain, but it also sends nerve impulses to several endocrine glands and to motor pathways governing the activity of skeletal muscle,

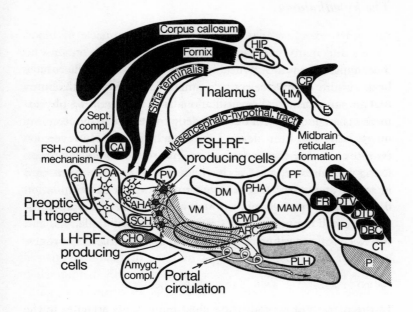

Fig. 2-1. Diagram of the hypothalamic control mechanisms for FSH and LH secretion. Black zones represent nerve tracts in the brain; outlined areas represent nuclei. Shaded portion indicates site of 'hypophysiotrophic area' containing the FSH-RF and LH-RF producing neurones whose axons are shown ending on the capillary loops of the hypophysial portal circulation. *Abbreviations used:* AHA, anterior hypothalamic area; Amygd. compl., amygdaloid complex; ARC, arcuate nucleus; CA, anterior commissure; CHO, optic chiasma; CP, posterior commissure; CT, nucleus centralis tegmenti; DBC, decussatio brachiorum conjunctivorum; DM, dorsomedial nucleus; DTD, decussatio tegmenti dorsalis; DTV, decussatio tegmenti ventralis; E, pineal gland (epiphysis); FD, fascia dentata; FLM, fasciculus longitudinalis medialis; FR, fasciculus retroflexus; GD, gyrus diagonalis; HIP, hippocampus; HM, medial habenular nucleus; IP, interpeduncular nucleus; MAM, mammillary nucleus; P, pons; PF, nucleus parafascicularis thalami; PHA, posterior hypothalamic area; PLH, posterior lobe of the hypophysis; PMD, dorsal premammillary nucleus; POA, preoptic area; PV, paraventricular nucleus; SCH, suprachiasmatic nucleus; Sept. compl., septal complex; VM, ventromedial nucleus. (From B. Flerko. In *The Hypothalamus*. Ed. L. Martini, M. Motta and F. Fraschini, pp. 351-63. London; Academic Press (1970).)

the heart, exocrine glands and the smooth muscle of blood vessels and many viscera. Yet the hypothalamus contains no mechanism that is indispensable for life. Hence researchers have been able to destroy selectively small foci in the hypothalamus and pursue exhaustive investigations into the resulting physiological changes. Much of our knowledge of its role in the economy of the body has been derived from such studies. However, inferences based on tissue destruction give only limited insights on living mechanisms, and for this reason attention has been paid in recent years to other techniques such as electrical stimulation of the intact hypothalamus through needle electrodes, implantation of small masses of sex hormone in the hypothalamus, and the electrical recording of signals from its constituent neurones.

HYPOTHALAMUS AND PITUITARY GLAND

In the context of reproduction most importance attaches to the relationship of the hypothalamus with the pituitary gland. The posterior part of the pituitary (the neurohypophysis) actually arises in development as a ventral outgrowth of the forebrain and is composed entirely of nervous elements and connective tissue cells (neuroglia). The nerve cell bodies are situated in the hypothalamus in the supraoptic and paraventricular nuclei (Fig. 2-2). They are rather special cells which, besides conducting electrical impulses, also manufacture the polypeptide hormones vasopressin and oxytocin. These are transported in the form of neurosecretory granules down the axons which pass via the pituitary stalk into the posterior lobe where the hormone is stored, and released in response to appropriate stimuli. Oxytocin is thought to be formed mainly by the cells of the paraventricular nucleus and there is very good evidence that it is released from the posterior lobe during suckling to cause ejection of milk from the mammary alveoli (see Chapter 5), and also in parturition where its effect is to speed up and strengthen the contractions of the uterus to aid expulsion of the fetus and placenta (Book 2, Chapter 3) (Fig. 2-3).

The anterior pituitary, which secretes a variety of hormones including FSH, LH and prolactin, is non-nervous in its embryological origin and arises as a dorsal evagination of the ectoderm of the roof of the buccal cavity (Rathke's pouch). Not long ago considerable controversy surrounded the question of the nervous regulation of the anterior pituitary, for although at the pituitary stalk the nervous and non-nervous portions of the organ are in close apposition (Fig. 2-2), morphological and experimental evidence established that nerve fibres do not penetrate to the secretory elements of the anterior pituitary in

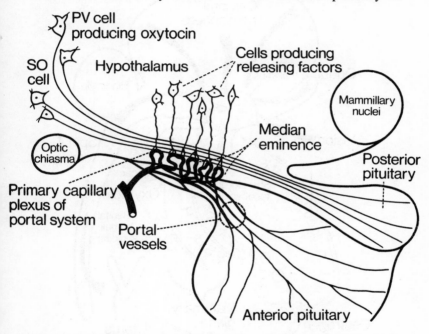

Fig. 2-2. Diagram to show relationship of nervous control of the posterior pituitary and neuro-humoral control of the anterior pituitary in the median eminence of the hypothalamus. The axons of the neuro-secretory cells of the paraventricular (PV) and supraoptic (SO) nuclei pass without interruption to the posterior lobe of the pituitary where their hormones are discharged into the general circulation. The anterior pituitary is controlled by releasing factors produced by neurones in the hypophysiotrophic area of the hypothalamus and discharged into the primary capillary plexus of the portal system. Compare with Fig. 2-1.

33

significant numbers, and direct innervation therefore plays no part in the normal functional activity of the gland. However, linking the median eminence of the hypothalamus with the anterior pituitary is the so-called hypophysial portal system of blood vessels (Fig. 2-2). From much painstaking research this vascular channel has been shown to convey the releasing factors secreted by hypothalamic neurones to the cells of the anterior pituitary. The activity of the anterior pituitary is thus subjected

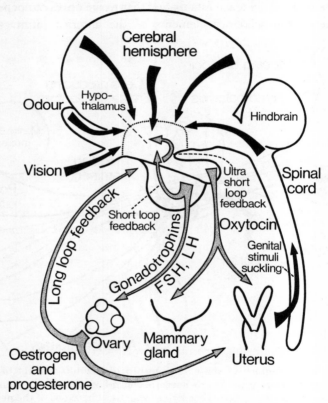

Fig. 2-3. This shows the various nervous inputs (dark arrows) and hormonal influences (pale arrows) impinging on the hypothalamus. The main target organs of the reproductive hormones from the pituitary (ovary, mammary gland and uterus) are also shown. Note the three postulated hormonal feed backs – long loop (from ovary), short loop (from anterior pituitary) and ultra-short loop (from hypothalamus itself).

to nervous control without the necessity for a direct innervation.

This distinction between the modes of functional attachment of the hypothalamus to the two lobes of the pituitary is responsible for the curious train of events that often follows surgical division of the pituitary stalk. Because most of the severed nerve fibres are destined for the posterior lobe, the secretory function of this gland virtually ceases, with the result that parturition may be difficult or impossible, or in the case of lactating animals milk removal by the suckling young may be greatly impaired – both effects due to deficiency of oxytocic hormone. The blood vessels of the hypophysial portal system that are severed, however, normally re-establish themselves within 2 or 3 days and the anterior pituitary can then resume normal working with the recurrence of oestrous cycles and ovulation. If the severed ends of the stalk are intentionally separated by an impenetrable barrier, the regrowth of vessels is prevented and in such animals all sexual functions are profoundly depressed, even to the extent of causing permanent anoestrus and sterility. Somewhat similar consequences ensue if the anterior pituitary is experimentally removed from its normal site in the skull, a pit called the sella turcica, and grafted elsewhere, e.g. under the kidney capsule. In this site the pituitary cells can be sustained by the vascularization acquired from the capsular vessels, but the releasing factors essential for stimulating the secretion of gonadotrophic hormones are deficient since they derive from the hypothalamus. Regrafting the pituitary back into the sella turcica, or directly into the hypophysiotrophic area of the hypothalamus (see Fig. 2-1), restores normal hormonal functioning. Prolactin, the hormone responsible for initiating milk secretion, is the exception to the rule, for this is formed in the renal grafts in greater amounts than occur in the intact pituitary gland.

HYPOTHALAMIC RELEASING FACTORS 1696562

When it first became clear that the anterior pituitary was regulated by humoral factors in the portal vessels, the natural expectation was that they might be similar or identical to other chemical substances known to be transmitters in nerve cells in other parts of the nervous system, e.g. acetylcholine, noradrenaline. All efforts, however, to establish a crucial role for these familiar transmitters proved abortive. Next, much work was done with extracts of the posterior lobe, which contained substances like vasopressin, which could stimulate anterior pituitary responses deceptively similar to those known to result from certain natural nervous stimuli. This approach also yielded disappointingly unconvincing results. Finally, as the result of long and exacting chemical procedures of extraction, purification, analysis and synthesis, there is good evidence that the releasing factors comprise a family of short chain polypeptides, some of them with lower molecular weights than the octapeptides of the posterior pituitary gland. There are about ten of these factors, and they may be chemically identical in different mammalian species. The two that are of most concern to us are FSH-RF and LH-RF. Some idea of potency of these substances can be gleaned from the fact that as much as 3 kg of pig hypothalamus may be needed to produce a milligram of LH-RF, but the minimum effective dose for a rat can be as low as 5×10^{-9}g. A few of the substances act not by stimulating release of their target pituitary hormone but by inhibiting release. The best known of these is the prolactin inhibiting factor (PIF). Supposedly owing to lack of this mediator, excess production of prolactin occurs in kidney grafts of pituitary tissue (see above).

The hypothalamic releasing and inhibiting factors all have to obey certain experimental criteria for recognition as true physiological entities. These include the demonstration of the pituitary response both *in vivo* and *in vitro* (tissue culture), and lack of effect in hypophysectomized animals. Rather an exciting trail of research has led from the original formulation of neurohumoral theory of anterior pituitary control via the portal vessels more than 20 years ago to the present state of knowledge. There are still important gaps, however, not the least of which is the cellular origin within the hypothalamus of the various releasing factors. They are presumed each to derive from separate neurones (neurosecretion) situated in the ventromedial or anteromedial hypothalamic areas (hypophysiotrophic area, Fig. 2-1), but none of the cells has been specifically identified as yet. Another problem, working back through the excitatory mechanism, is the identification of the synaptic transmitter that induces release of the releasing factor, i.e. by exciting say the LH-RF producing neurones. A currently fashionable hypothesis is that dopamine, whose presence in the median eminence region has been shown by fluorescence microscopy, is the synaptic transmitter at this site. To carry conviction, such ideas will need to be examined in depth by sophisticated electrophysiological and neuropharmacological techniques. At present these are only possible for the large neurones of the supraoptic and paraventricular nuclei (Fig. 2-2), which are identifiable in the living animal by stimulating their axons in the posterior pituitary gland with electric shocks and recording with micro-electrodes the resulting back-fired nerve impulse invading the cell body. On these cells there seem to be both excitatory and inhibitory synaptic endings, which probably employ acetylcholine and noradrenaline as their respective transmitters. There is certainly no assurance that the mechanisms controlling releasing factors will prove less complicated than this. Indeed they may well turn out to be even more complex, if only because there are several releasing factors and only two posterior pituitary hormones.

CONTROL OF OESTROUS CYCLES

Without doubt the laboratory rat is the best understood animal, so far as the hypothalamic mechanisms regulating the sexual cycle are concerned. We have already seen that cycles cease if the pituitary is functionally disconnected from the hypothalamus. But it is not necessary to cut the pituitary stalk to do this. Severing the connections between the median eminence and the anterior hypothalamus and preoptic region is sufficient, either by circumscribing knife cuts (small hypothalamic islands) or

The hypothalamus

simply by transverse cuts in front of the median eminence. This implies that nervous inputs from anterior 'centres' are essential for normal cyclic activity. One of the most important discoveries in reproductive physiology in the last 30 years is the so-called 'critical period' which in the rat usually lasts only about 2 hours on the afternoon of pro-oestrus. During this period excitatory signals are generated in the hypothalamus that release LH-RF in sufficient amounts to evoke the discharge of an ovulatory quota of pituitary gonadotrophin. Administration at this time of the barbiturate Nembutal or many other anaesthetic agents blocks ovulation, but appropriate stimulation of the pre-optic area in Nembutal-blocked rats can induce LH-RF secretion and restore the normal ovulation.

In common parlance this phenomenon is sometimes referred to as the hypothalamic 'clock'. In the rat the alarm bell rings at a prescribed hour every 4 (or 5) days. There is no doubt that the diurnal alternation of light and dark plays an integral part in the mechanism, for continuous lighting quickly stops the rat oestrous cycle. Moreover, when Nembutal is given to a cyclic female rat during the critical period, ovulation is not just set back by the few hours the animal is narcotized, but is delayed by a full day. Many bodily functions, including a number of endocrine activities, are known to undergo a circadian (approximately 24-hour) rhythm and the hypothalamus is no exception. But why a 4-day clock? The ovary, with its hormone output acting back onto the preoptic and anterior hypothalamic mechanisms, seems to provide the answer. If the oestrogen secreted by the ovary early in pro-oestrus is inactivated either by antibodies or by synthetic anti-oestrogens, the effect is to block LH-RF release, LH secretion and the resulting ovulation. In other words ovarian oestrogen seems to be the signal to the hypothalamus to initiate LH-RF secretion during the critical period. This idea fits with fairly extensive evidence of selective uptake of labelled oestrogen by cells in the anterior hypothalamus. Now, the firing rate of the neurones can be recorded with microelectrodes inserted into the hypothalamus.

38

When this is done in groups of rats taken at different stages of the oestrous cycle, a peak of activity is found to occur on the afternoon of pro-oestrus in the preoptic and anterior hypothalamic region but not elsewhere, and not on any other day of the cycle. Moreover this cyclical change can be seen in rats with large hypothalamic islands, where there is no possibility of nervous inputs from other regions of the CNS, and is prevented by removal of the ovaries early on the day of dioestrus, i.e. before the oestrogen is secreted. It is very tempting therefore to think that we are seeing in these neuronal activities the actual working parts of the hypothalamic ovulation clock. But experience teaches that nature is seldom so generous, and a good deal more tinkering will probably be necessary before the purported mechanism conforms to reality. For one thing the clock can be stopped, e.g. by copulation, when cycles are temporarily suspended by the pseudopregnant interval, or by conception and pregnancy, or by suckling. Perhaps the most practically advantageous disturbance of the ovulatory clock mechanism is that produced by the contraceptive pill. This comprises one or more progestagenic agents whose principal mode of action seems to be to inhibit the hypothalamic trigger for LH secretion.

The hypothalamic mechanisms of all these processes remain obscure. In the modern idiom explanations are usually sought in terms of positive or negative hormonal 'feedbacks', which may be 'long-loop', i.e. via the ovaries, 'short-loop', i.e. via the pituitary, or 'ultra-short-loop', i.e. humoral feedback direct from the hypothalamus (Fig. 2-3). Also invoked are afferent nervous inputs from various other brain regions. Evidence for many of these notions is rather inconclusive, and they will not be pursued further here, except to acknowledge that mechanisms assuredly exist within the hypothalamus to integrate nervous and hormonal signals separated both in space and time.

HYPOTHALAMUS AND SEXUAL BEHAVIOUR

In 'spontaneously' ovulating mammals the behavioural cycle of mating activity is synchronized with the ovulatory cycle, with the usual effect that spermatozoa are available in the oviduct before the shedding of the eggs from the ovary. These behavioural responses also depend upon the hypothalamus. One of the most dramatic illustrations of this is the fact that small implants of crystalline oestrogen placed in the anterior hypothalamus of spayed cats can in a few days turn an aggressively frigid female into a raging nymphomaniac, even though her own genital tract is atrophic from the lack of circulating ovarian hormones. But it is not simply a question of the sex behaviour being determined by the 'gender' of the sex hormones to which it is exposed, for we now know that the hypothalamus is itself programmed for post-pubertal sexual function at a very early stage of development, i.e. in the perinatal period. The female pattern of cyclical activity, both in the genital tract and in behaviour, appears to be the basic programme and will manifest itself in the post-pubertal stage unless the differentiating nervous tissue is exposed in the neonatal period to androgenic hormone – either from the developing testis or from external administration. Thus, for example, if the newborn male rat is castrated, its hypothalamus develops in the female pattern, with the result that if a sister's ovaries and vagina are grafted into the animal they undergo the usual cycles of follicular growth and vaginal cornification. Conversely, a newborn female that is injected with androgen, when it reaches maturity remains acyclic, anovular and displays homosexual behavioural tendencies. The evidence that similar transexual changes in the hypothalamus occur in man is still inconclusive but it has been reported that girls born of mothers treated during pregnancy with progestagens having androgenic side effects may develop as tomboys, disdaining the normal girlish delights of dolls and dresses. ('Brain sex' is discussed in more detail in Book 2, Chapter 2.)

The scientific literature on the hypothalamus has snowballed incredibly in the last 20 years – not only because the hypothalamus appears to be the cerebral vehicle for all the more colourful 'vices' – eating, drinking, sleeping and sex – but also because of the intellectual challenge presented by its inaccessibility and comparatively innocent morphology. That so much of our knowledge is confined to animal species such as the laboratory rat is to be regretted. If we knew more about human hypothalamic mechanisms we might better deal, not only with contraceptive and fertility problems, but also with a variety of psychosexual conditions, e.g. premenstrual tension, impotence, homosexuality and sexual psychopathy. What gains in human happiness might then ensue!

SUGGESTED FURTHER READING

Recent studies on the hypothalamus. Ed. K. Brown-Grant and B. A. Cross. *British Medical Bulletin* 22, no. 3 (1966).

Hypothalamic Control of the Anterior Pituitary. J. Szentagothai, B. Flerko, B. Mess and B. Halàsz. Budapest; Akademiai Kiado (1968).

Frontiers in Neuroendocrinology. Ed. W. F. Ganong and L. Martini. London; Oxford University Press (1969).

The Hypothalamus. Ed. W. Haymaker, E. Anderson and W. J. H. Nauta. Springfield; Thomas (1969).

The Hypothalamus. Ed. L. Martini, M. Motta and F. Fraschini. London; Academic Press (1970).

Mammalian Neuroendocrinology. B. T. Donovan. London; McGraw-Hill (1970).

3 Role of hormones in sex cycles
R. V. Short

In the first two chapters of this book, we have learned something of the baffling array of hormones and other biologically active substances that are produced by the hypothalamus, the pituitary gland, the gonads, and the reproductive tract itself. We must now try to fit all these pieces of the jigsaw into an integrated picture of the reproductive cycle in the male and female mammal. This is not a simple task, for not only are we faced with major differences between the sexes, but we have the added complication of species differences which are often so great that generalizations become impossible. In the male, we must distinguish between continuous breeders and seasonal breeders, and in the female, between spontaneous and induced ovulators, oestrous and menstrual cycles. The position in the female is made even more complex by the fact that the ovary is not a homogeneous organ, but an aggregation of two or three distinctly different endocrine tissues that wax and wane during the sexual cycle. So it is not surprising that people find this a challenging subject to master.

THE MALE

In the present state of our knowledge, the hormonal control of reproduction in the male is relatively simple. But one has the uneasy feeling that this is only because it has not been investigated thoroughly. Testicular activity can be divided into its sperm-producing and hormone-producing components, and we must consider them separately.

SPERMATOGENESIS

The production of spermatozoa is under the control of only two of the pituitary hormones, FSH (follicle stimulating hormone), and LH (luteinizing hormone), which is also referred to in the male as ICSH (interstitial cell stimulating hormone). If the pituitary gland is removed from a male rat, the testes rapidly regress, and spermatogenesis comes to a halt (see Fig. 3-1).

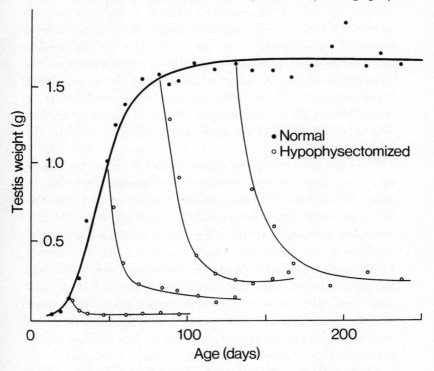

Fig. 3-1. Decline in testicular weight of rats following hypophysectomy at different ages. (From R. Ortavant, and M. Courot. *La physiologie de la reproduction chez les mammifères*. p. 111, Fig 1. Masson et Cie; Paris (1968).)

LH seems to be the more important of the two pituitary hormones in the adult, since some degree of spermatogenesis can be maintained in hypophysectomized rats if they are given LH alone as replacement therapy following the operation. But the mode of action of LH in maintaining spermatogenesis is quite complex, since it has a number of different sites of action. In

43

addition to a direct effect on the germinal epithelium, it also exerts an indirect effect by way of testosterone. We know that LH stimulates testosterone secretion by the Leydig or interstitial cells that surround the testis tubules, and although much of this testosterone escapes by way of the spermatic veins and lymphatics to reach the general circulation, a significant amount probably filters through into the lumen of the testis tubules where it may have an important local action on the seminiferous epithelium. LH also seems to have a direct effect on the Sertoli cells within the tubules. These cells may be capable of some degree of steroid synthesis, and they certainly act as nurse cells for the developing spermatozoa. LH may control the rate at which they release their infant spermatozoa into the lumen.

FSH seems to act synergistically with LH, since the effects of the two hormones in combination are greater than those of either in isolation. FSH may also be more important than LH in the development of the gonad before puberty, with LH only assuming the dominant role after spermatogenesis has become established. FSH does not seem to influence steroid secretion by the testis, so that its effects are confined to a direct action on the germinal epithelium and also probably on the Sertoli cells. Only now, with the development of tissue culture techniques, and the availability of highly purified pituitary hormones and specific antisera raised against them, are we in a position to give this subject the renewed critical interest that it deserves.

Most endocrine systems in the body have a built-in safety device, the negative feedback system, which allows the product to regulate the amount of hormone that is released. There has been much speculation in the past about whether the pituitary is aware of the number of spermatozoa that the testis is producing, and whether it is able to modulate its gonadotrophin output accordingly. For example, if one testis is removed from a young animal, the remaining one will undergo a compensatory enlargement. This has suggested to some people that there might be another testicular hormone, provisionally named 'inhibin', whose production is a function of the number of spermatozoa produced, and whose action is to inhibit pituitary gonadotrophin secretion. But nobody has yet been able to isolate inhibin, and the effects of unilateral castration can equally well be explained by the known feedback effects of testosterone on LH production.

STEROIDOGENESIS

Turning to the endocrine activity of the testis, the situation at first sight seems even simpler. We have but one pituitary hormone, LH, and one principal testicular hormone, testosterone, to consider. Injections of LH can stimulate testosterone production, even in the prepubertal animal which would normally secrete only a trace amount. A convenient way of assessing androgen secretion indirectly is to measure the levels of fructose and citric acid in the ejaculate, these two compounds being secreted by the seminal vesicles under the influence of testosterone (see Fig. 3-2). We also know that there is a functional negative feedback system, for testosterone, progesterone or oestrogen injections will depress the blood LH levels, and castration raises them (see Fig. 3-3). Unlike testosterone, oestrogen seems to depress FSH as well as LH secretion, and hence it may seriously interfere with spermatogenesis.

As we shall see later, the feedback effects of gonadal steroids on the hypothalamus are quite different in the male and female; whereas oestrogen is always *inhibitory* in the male, it can have a *positive* feedback effect in the female, and stimulate the discharge of LH. This basic difference between the sexes may be the result of hypothalamic imprinting by androgens during fetal or neonatal life; it could account for the relatively constant pattern of gonadotrophin release in the adult male, and the cyclical pattern in the female that is responsible for oestrous or menstrual cycles. This is discussed further in Chapter 2, Book 2.

But just how constant are the gonadotrophin levels in males, and how effective is the negative feedback control? Studies of men from whom blood samples have been taken hourly for

24 hours show that the blood levels of FSH are relatively stable, although there may be fluctuations in the blood LH and testosterone levels, particularly during sleep. But in bulls and rams, there are enormous fluctuations in LH secretion throughout the 24 hours, and this intermittent stimulation of the testis results in a testosterone secretion that is continuously vacillating between basal and maximal levels during the course of the day (see Fig. 3-4). In these animals, the negative feedback mechanism seems seldom to be called into play, and it does not normally regulate the periodic spontaneous bursts of LH secretion from the pituitary.

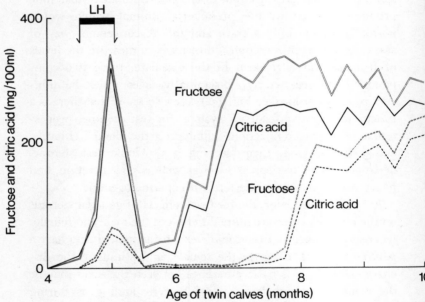

Fig. 3-2. Effects of LH treatment on the testicular activity of identical twin bull calves before puberty. The animal on a high plane of nutrition (solid lines) showed a pronounced increase in the fructose and citric acid levels in the ejaculate in response to the treatment, and normal puberty occurred at about 7 months of age. The animal on a low plane of nutrition (dotted lines) showed a poor response to the same dose of LH, and normal puberty was delayed by about 2 months. (From D. V. Davies, T. Mann, and L. E. A. Rowson, *Proc. Roy. Soc. B* **147,** 332, Fig. 4 (1957).)

We are beginning to realize that sexual excitement may provoke a discharge of LH from the pituitary, and so stimulate a secretion of testosterone by the testis, although whether this has any functional significance we cannot say; it could facilitate the release of spermatozoa from the Sertoli cells into the lumen

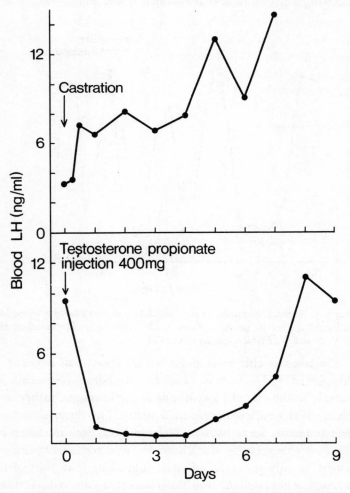

Fig. 3-3. Effects of castration and testosterone injection on the blood LH level in rams.

of the seminiferous tubules. Figure 3-5 shows the sudden release of LH in a bull when he was teased by being led up to a cow but prevented from serving her. That a similar mechanism may be at work in man is suggested by the fact that the rate of beard growth may increase in anticipation of female company, following a period of sexual abstinence (see Fig. 3-6).

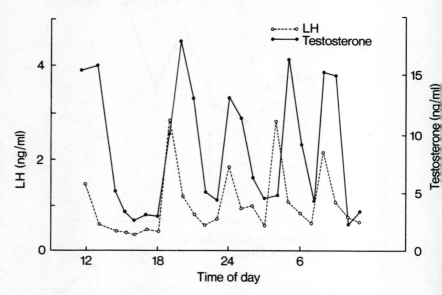

Fig. 3-4. Normal fluctuations in blood LH and testosterone levels in a bull during a 24-hr period. (From C. B. Katongole, F. Naftolin, and R. V. Short. *J. Endocr.* **50,** 457 (1971).)

The testis is able to respond within about half an hour to changes in LH secretion, and this response represents an increase in the rate of biosynthesis of the hormone, rather than release of steroid from preformed stores. The turnover time of testosterone in the testis, by which we mean the time taken for the testis to secrete as much hormone as it contains at any one instant, is only about 10 minutes, suggesting that most of the hormone is liberated into the blood almost as soon as it is formed.

We said earlier that testosterone was the principal steroid secreted by the testis, but this is not strictly correct. The testis also secretes oestrogen, and in some animals, notably the stallion and boar, enormous amounts are produced; the concentrations excreted in urine far exceed those in the urine of pregnant mares or sows. We have no idea of the functional significance of all this oestrogen, although much of it seems to be secreted in a conjugated form, which is relatively inactive biologically. This is

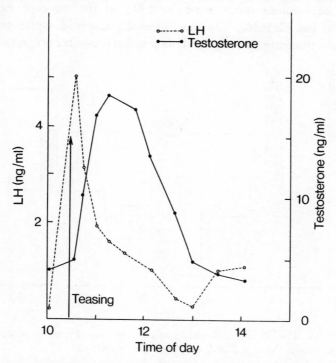

Fig. 3-5. The effect of sexual excitement on the blood LH and testosterone levels in a bull. (From C. B. Katongole, F. Naftolin, and R. V. Short. *J. Endocr.* **50**, 457 (1971).)

probably just as well, since oestrogen injections are an effective way of suppressing spermatogenesis in many species.

Another testicular steroid of some interest is androstenedione (see Chapter 1). Although it is a very weak androgen, it is the major secretory product of the testis prior to puberty in the bull. At the time of puberty, the pattern of secretion changes

Role of hormones in sex cycles

abruptly in favour of testosterone. For a time therefore the cause of puberty seemed to be related to changes in the activity of the testicular enzyme system (17β-hydroxysteroid dehydrogenase) that normally converts androstenedione into testosterone, but studies of testicular steroid secretion before and after puberty in other species have failed to show any such androstenedione–testosterone 'switch', so the concept now seems less plausible. The most generally accepted explanation for the phenomenon of puberty is that it represents a progressive

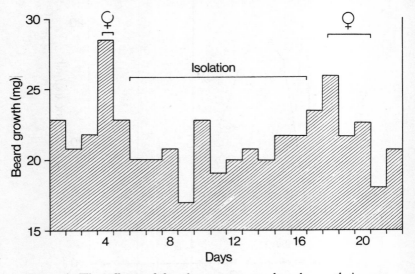

Fig. 3-6. The effects of female company on beard growth in man. (From Anon. *Nature, Lond.* **226,** 869, Fig 1 (1970).)

loss of hypothalamic sensitivity to the inhibitory effects of steroidal feedback. Before puberty, minute amounts of gonadal steroid are able to inhibit the hypothalamus and so prevent pituitary gonadotrophin release, but as the hypothalamus gradually escapes from this inhibition, so the pituitary is enabled to secrete more and more gonadotrophins until eventually spermatogenesis is complete and there is sufficient androgen to stimulate the development of male secondary characteristics. This simple concept of puberty may have to be modified soon, because recent studies have shown that the FSH levels in male rats are highest in prepubertal animals.

In many wild animals, the males are seasonal breeders that lapse into a period of complete sexual quiescence for part of the year; the testes regress, spermatogenesis comes to a halt, and androgen secretion declines to a point where it is no longer adequate to maintain the secondary sexual characteristics. Deer are a case in point, and could be said to undergo an annual 'puberty' and 'senescence' whose timing is determined by daylight length.

EFFECTS OF ANDROGENS ON TARGET ORGANS

A few years ago people regarded the male reproductive tract as the principal target organ for testosterone; it was well recognized that androgens controlled the size and secretory activity of the male accessory organs, such as the seminal vesicles, prostate and bulbo-urethral glands. But now we are gradually beginning to appreciate that androgens have effects in most cells in the body.

In addition to their regulatory action on the hypothalamus, androgens are essential for the development and maintenance of male sexual behaviour. They may also control certain aspects of social behaviour, since they can increase an individual's aggression, and hence his position in a social hierarchy. Loss of hair from the scalp, and growth of hair on the chin, are two well-known androgenic effects in man, but less well known are the stimulatory effects of androgens on sebaceous glands, possibly accounting for the acne of pubertal youths.

In some species, such as mice and pigs, androgens may have distinct effects on the salivary glands, either altering their histological appearance, or in the case of the boar, causing a pheromone (olfactory hormone) to be secreted which excites the female. The growth and shedding of antlers in male deer is perhaps one of the more picturesque examples of an androgen effect, as is the brilliant plumage of many male birds. The action of androgens on the larynx, causing 'breaking of the voice', probably occurs in some animals as well as in man. The development of certain enzyme systems in the liver, and β-glucuronidase and alcohol dehydrogenase in the kidney, are yet other effects. In the cat family, from lions to tomcats, the characteristic offensive odour of the male is caused by a lipid secreted by special cells in the kidney under the influence of testosterone.

The anabolic or muscle-building effects of androgens are well-recognized, and made use of clinically to aid the recovery of debilitated patients whose muscles have wasted away, or used illegally to improve the performance of athletes. The reason why we still continue to castrate domestic animals destined for the table is probably related more to the change it produces in their behaviour than to any beneficial effects on growth rate, although in pigs the intact male produces this pheramonal steroid whose odour taints all the meat and is particularly offensive to women.

When one realizes that even a dog's propensity to cock his leg is androgen-dependent, one begins to wonder where the story will end. But it would be wrong to think that a man without his male hormone is useless, because we know that historically eunuchs have played an important role in some civilizations; they frequently became the trusted confidants of kings and rulers, and their advice was sought on all matters of state. Eunuchs also became eminent philosophers and politicians; curiously, one even rose to fame as a general under Justinian, which is hardly what one would expect of a man deprived of his aggression-inducing hormone.

THE FEMALE

Unfortunately, we cannot describe the reproductive cycle of a 'typical' female animal, and assume that the one explanation will hold true for all mammals. Although the same pituitary and ovarian hormones are common to most species, the way in which they are controlled and the effects they produce vary from one to another. In the discussion that follows, we will describe the sequence of events in the oestrous cycle of sheep and the menstrual cycle of women, and see how these differ from the short oestrous cycles of the rat.

THE OESTROUS CYCLE OF THE EWE

In the Northern hemisphere, sheep are seasonal breeders with regular 16-day oestrous cycles that start in the autumn and cease again in the spring; the principal environmental stimulus that controls reproductive activity seems to be daylight length. The ewe will show oestrous behaviour for only about 1 day out of the 16, and ovulation occurs about 24 hours after the onset of oestrus. It is customary to call the day of onset of oestrus Day 0 of the cycle, but this is not the day on which all the exciting hormone changes occur. Because of the considerable time lags between the onset of pituitary gonadotrophin secretion and an ovarian response, and then between the onset of ovarian steroid secretion and a behavioural response, the hormone changes that produce oestrus precede it by a day or two.

The corpus luteum will have been secreting progesterone maximally from about day 7 of the previous cycle, and then on day 15, a day before the animal starts to show oestrus, progesterone secretion declines abruptly (see Fig. 3-7). This is the event that triggers off the whole sequence of hormone changes that produce the next oestrus and ovulation. If the corpus luteum is made to regress prematurely, the animal returns to oestrus sooner than expected; if the life of the corpus luteum is prolonged, oestrus is postponed. Therefore the functional corpus luteum is the time-clock that controls the length of the oestrous cycle.

Many years ago, Leo Loeb in America made the curious observation that removal of a guinea pig's uterus would prolong the length of life of the corpora lutea. This observation was subsequently confirmed, and found to be true also of sheep and other domestic animals. Careful studies by numerous groups of

Role of hormones in sex cycles

workers, including du Mesnil du Buisson in France, Bob Melampy and his group in Iowa, and Tim Rowson and Bob Moor in Cambridge, showed that the uterus exerted a 'lytic' action on the corpus luteum. Removal of this lytic influence by hysterectomy (taking out the uterus) prolongs the life of corpora lutea. The lytic influence seems to operate predominantly in a local manner, because if the ovaries are removed from the vicinity of the uterus and transplanted elsewhere in the body, oestrous cycles are prolonged. Similarly, if one horn of the uterus is removed, the life of corpora lutea on that side will be prolonged, whereas corpora lutea on the side of the intact uterine

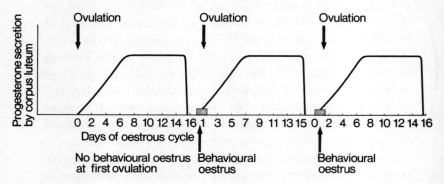

Fig. 3-7. The oestrous cycle of the ewe, showing how the first ovulation of the season is unaccompanied by oestrus. Note short time interval between regression of the corpus luteum and the next ovulation.

horn regress normally (see Fig. 3-8). The timing of the lytic stimulus seems to depend on the length of time that the endometrium has been under the influence of progesterone, because if sheep are placed on continuous progesterone therapy and then made to ovulate in response to gonadotrophin injections, the corpora lutea will regress almost as soon as they are formed.

So it seems that there is a local utero-ovarian cycle, whereby the corpus luteum stimulates the uterus to produce a substance which in turn destroys the corpus luteum (see Fig. 3-9). This lytic substance is certainly formed in the endometrium rather than the myometrium, and there is growing evidence that it is

54

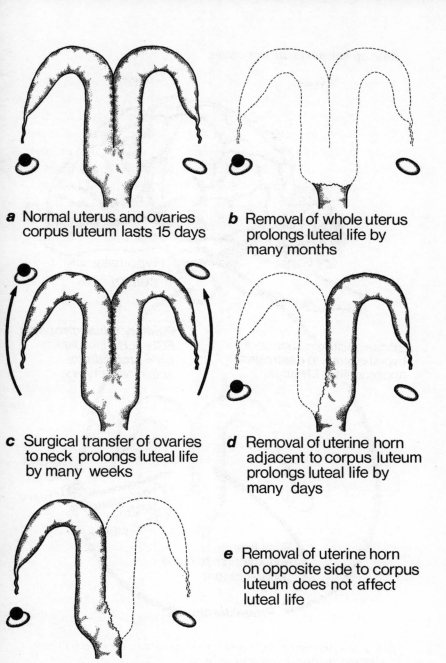

a Normal uterus and ovaries corpus luteum lasts 15 days

b Removal of whole uterus prolongs luteal life by many months

c Surgical transfer of ovaries to neck prolongs luteal life by many weeks

d Removal of uterine horn adjacent to corpus luteum prolongs luteal life by many days

e Removal of uterine horn on opposite side to corpus luteum does not affect luteal life

Fig. 3-8. The local lytic effect of the uterus on the corpus luteum in sheep. For the sake of simplicity, the corpus luteum is always shown in the left ovary.

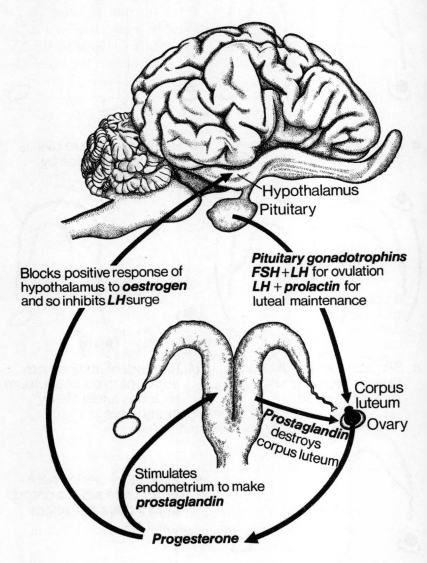

Hypothalamus

Pituitary

Blocks positive response of
hypothalamus to **oestrogen**
and so inhibits **LH** surge

**Pituitary gonadotrophins
FSH+LH** for ovulation
LH + prolactin for
luteal maintenance

Corpus
luteum

Ovary

Prostaglandin
destroys
corpus luteum

Stimulates
endometrium to make
prostaglandin

Progesterone

Fig. 3-9. Interrelationships between the pituitary, the corpus luteum,
the uterus and the hypothalamus in sheep.

56

prostaglandin $F_{2\alpha}$. But the local action is still difficult to understand. We do know that prostaglandin is largely destroyed in one passage through the lungs, which would minimize its systemic effects; in some mysterious way prostaglandin seems able to diffuse from the uterine vein directly into the ovarian artery (see Fig. 3-10). Infusions of prostaglandin $F_{2\alpha}$ into either the uterine vein or the ovarian artery stop progesterone secretion

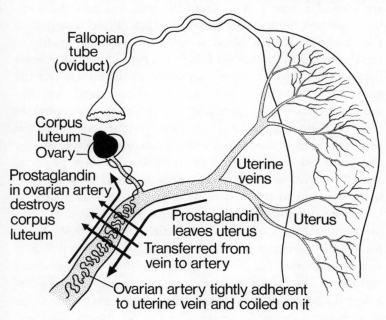

Fig. 3-10. Postulated route by which prostaglandin manufactured by the progesterone-primed uterus is able to enter the ovarian artery and destroy the corpus luteum in sheep.

more effectively than if they are given into the peripheral circulation, and prostaglandin is present in uterine venous blood in high concentrations on day 15 of the cycle.

Pharmacological doses of oestrogen given to ewes early in the cycle can prolong luteal life, whereas if given later they hasten luteal regression. The oestrogen seems to be acting primarily on the uterus, and in doing so interferes with the

57

production of prostaglandin by the endometrium, and thus affects the life of the corpus luteum.

If the corpus luteum is the time-clock of the oestrous cycle, we must examine more closely the physiological factors that control this transitory endocrine gland, which is here today and gone tomorrow; we must consider the factors responsible both for its formation, and for its regression. Luteal cells in the sheep are derived almost entirely from the granulosa cells that form the innermost lining of the Graafian follicle. Prior to ovulation, they are starved of a blood supply, and virtually unable to secrete steroids. But once the follicle has ruptured, blood vessels break through the basement membrane of the follicle and the granulosa cells begin to luteinize. By this we mean that they increase in numbers (hyperplasia) and in size (hypertrophy), as the cytoplasm expands and a smooth endoplasmic reticulum begins to develop. Coincident with these morphological changes, the granulosa cells begin steroid synthesis. The principal steroid secreted is progesterone, because luteal cells are relatively deficient in the enzyme systems that transform progesterone into androgens and oestrogens (see Chapter 1). The theca interna cells of the intact follicle, on the other hand, possess these enzyme systems to the full, so that their principal secretory product is not progesterone but androgen and oestrogen. The transition between an oestrogen-secreting follicle and a progesterone-secreting corpus luteum is therefore due to a change in the predominant endocrine cell type in the ovary, which is in turn a natural consequence of follicular rupture.

If the corpus luteum is to secrete steroids at all, it needs the support of luteotrophic hormones from the anterior pituitary, which act in two ways. LH enters the follicular fluid of the pre-ovulatory follicle, and apparently has a long-term 'programming' action on the granulosa cells, that determines their level of secretory activity after ovulation. This pre-ovulatory luteotrophic stimulus is reinforced by the continuous secretion of luteotrophic hormones from the pituitary during the luteal phase of the cycle. If sheep are hypophysectomized soon after ovulation, progesterone secretion by the corpus luteum will stop in the next 5–10 days; this can be prevented by giving a mixture of prolactin and LH. Both hormones are relatively ineffective if given by themselves, so it seems preferable to refer to a 'luteotrophic complex', rather than to any particular gonadotrophin as *the* luteotrophic hormone. Although LH can cause a transitory increase in progesterone secretion by the corpus luteum, LH levels are low and do not fluctuate during the luteal phase of the cycle, when the corpus luteum is probably secreting at about maximal capacity in any case. Progesterone could indeed be regarded as the all-or-nothing hormone, which is either secreted maximally, or not at all.

Turning once more to the factors that control the regression of the corpus luteum, we have already seen how it is normally 'murdered' by the luteolytic action of the uterus at the end of the oestrous cycle. Since the blood levels of LH and prolactin are unaltered during luteal regression, this is clear evidence that the lytic stimulus is dominant to any pituitary trophic one. In the next chapter, Brian Heap discusses how the presence of an embryo in the uterus can prevent this 'murder' taking place, thus allowing the corpus luteum to persist and maintain the pregnancy.

Although one might imagine that the easiest way of destroying the corpus luteum would be by 'starving' it of pituitary gonadotrophic support, this is a way that Nature never seems to use, at least in the sheep. Perhaps this is because when luteal regression takes place, it needs to be abrupt, whereas death from starvation, as is seen following hypophysectomy, tends to be a rather long drawn out procedure. In addition to 'murder' and 'starvation', a corpus luteum can die of 'old age', and this is the fate that seems to befall it in the hysterectomized animal, where it will persist for about 5 months before slowly dwindling away.

Armed with a clearer picture of the formation and fate of the corpus luteum, we may now consider its function. The principal peripheral actions of progesterone are undoubtedly to stimulate

59

the development of the endometrial glands, inhibit the contractility of the uterine muscle, and develop the mammary
gland. But what of the central effects of the hormone?

We have already mentioned that the key property of a 'female'
hypothalamus is its ability to respond to oestrogen by initiating
a massive discharge of LH from the pituitary. However, the
reflex is blocked by progesterone, and this must be regarded as
one of the most important functions of the corpus luteum. It
explains why the periodic waves of follicular development and
oestrogen secretion that occur during the luteal phase of the
cycle in sheep cannot result in ovulations, so that the developing
follicles at this time are doomed to become atretic.

A second most important central action of progesterone is
related to sexual behaviour. If an ovariectomized ewe is injected
with a physiological dose of oestrogen (10 micrograms; 10^{-5}g),
she will not come into oestrus. However, if she has been given
an injection of progesterone a day or two beforehand, then she
will show oestrus. The requirements for normal oestrous
behaviour therefore seem to be a declining level of progesterone,
rapidly followed by a rising level of oestrogen. This explains
why most ewes fail to show behavioural oestrus at the first
ovulation of the breeding season (see Fig. 3-7).

So far we have just considered progesterone, and you might
be beginning to think that it is the only hormone of any significance in the female reproductive cycle. But whilst it makes
many of the major decisions about the timing of the cycle, we
must not neglect the vital role of the pituitary gonadotrophins.
Immediately, we run into difficulties, because our knowledge is
still in a state of flux.

The classical story is that FSH causes follicular development,
and LH stimulates steroid secretion by the developing follicle
and ultimately brings about ovulation. But we do not really
understand why the follicle should suddenly start secreting
oestrogens within a few hours of the regression of the corpus
luteum. Although it would be tempting to imagine that progesterone had been inhibiting the production of FSH by the

pituitary, blood FSH levels do not seem to rise immediately after luteal regression. And in any case, the ovary normally takes days rather than hours to respond to injections of FSH. Similarly, there is no change in the blood LH levels at this time. Maybe the gonadotrophin levels do not change, but there is an alteration in the sensitivity of the ovary to pre-existing levels. Clearly, there is a gap in our understanding, and all we can say is that somehow the regression of the corpus luteum stimulates the development of a Graafian follicle and the secretion of oestrogen.

The hypertrophied appearance of the theca interna cells gives a clue to where the oestrogen is coming from, and although much escapes into the ovarian vein, some gets through into the follicular fluid where it may stimulate the development of the granulosa cells. The rising oestrogen levels, in the absence of progesterone, trigger off the ovulatory discharge of LH from the pituitary after a time lag of about 12 hours (see Fig. 3-11), and it is this pre-ovulatory surge of LH that may account for the sudden spurt of follicular growth. LH seems to increase ovarian blood supply and decrease the permeability of the blood–follicle barrier, so assisting tissue fluid to penetrate the basement membrane and contribute to the follicular fluid; it is also responsible for initiating the resumption of meiosis by the oocyte.

A curious fact is that the principal steroid secreted by the ovaries at oestrus is not oestradiol-17β, but the weakly active androgen, androstenedione. This androgen may be extremely important in eliciting certain components of oestrous behaviour, a topic that is discussed in detail by Joe Herbert in Book 4, Chapter 2.

We still do not know what sequence of anatomical and biochemical events actually causes ovulation (see Book 1, Chapter 2), but there is a time lag of almost 24 hours between the LH surge and the moment of follicular rupture. During this intervening period, the oestrogen levels, having reached their peak, have started to fall again, so that oestrogen secretion may almost have ceased by the time ovulation actually takes place (see Fig. 3-11). Why this should be so is something of a mystery, but it may give added significance to the ovulatory release of oestrogen trapped in the follicular fluid, which will suddenly come into contact with the fimbria and Fallopian tube and may influence the collection and transport of the ovulated egg.

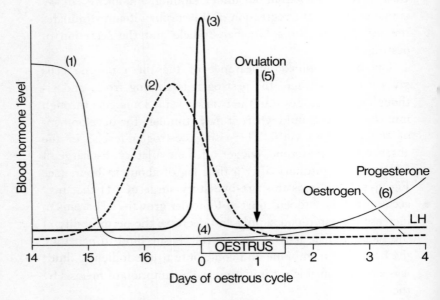

Fig. 3-11. Sequence of hormone changes leading up to oestrus and ovulation in sheep. Key to numbers in text (p. 63).

There is a considerable time-lag between the beginning of oestrogen secretion and the onset of oestrous behaviour, so that oestrogen secretion is already falling before the ewe shows the first signs of coming into heat, and oestrous behaviour will continue long after the hormone has almost vanished from the circulation.

Although nobody has yet found major changes in the blood FSH levels around oestrus, there is a big increase in prolactin secretion, which may be related to the future secretory activity of the corpus luteum. Following ovulation, there is a steady secretion of small amounts of prolactin and LH throughout the luteal phase, which are probably vital for the wellbeing of the corpus luteum.

Before we finally leave the topic of the oestrous cycle of the ewe, it may be helpful to summarize what has been said by listing the sequence of events depicted in Fig. 3-11.

(1) The corpus luteum regresses abruptly on day 15 owing to the lytic action of prostaglandin $F_{2\alpha}$, formed in the endometrium under the influence of progesterone and reaching the ovary by a local route.

(2) As a result of the falling progesterone levels, a Graafian follicle begins to secrete significant amounts of androgen and oestrogen, which reach peak values within about 24 hours.

(3) The rising oestrogen secretion triggers off a surge of LH release from the pituitary after a time-lag of about 12 hours. This LH surge can only be induced by oestrogen in the absence of progesterone. In addition to causing ovulation itself, some LH enters the follicular fluid and acts on the granulosa cells, programming their future secretory activity. The LH may also cause follicular enlargement, by increasing the permeability of the blood–follicle barrier, and it initiates the resumption of meiosis by the oocyte.

(4) The high oestrogen and androgen levels eventually bring the animal into behavioural oestrus, by which time the oestrogen secretion is already falling again.

(5) Ovulation occurs about 24 hours after the onset of the LH peak. The granulosa cells can now acquire a blood supply and begin to secrete progesterone.

(6) As the corpus luteum develops, so the progesterone levels begin to rise. The corpus luteum gradually acquires an increasing dependence on the constant low secretion of pituitary LH and prolactin for its maintenance and secretory activity. The high levels of progesterone in the luteal phase prevent any sudden surges of LH release in response to the occasional waves of follicular development and oestrogen secretion that occur.

A similar sequence of events occurs during the oestrous cycle of the cow and pig, except that the timings are quite different.

THE HUMAN MENSTRUAL CYCLE

The most obvious difference between the oestrous cycle of the ewe and the human menstrual cycle is the outward and visible sign. Women do not show periodic oestrus, which mercifully for society seems to have been completely suppressed during the course of human evolution, but instead they have a blood-stained vaginal discharge, the menses. As the name implies, menstruation occurs about once a month, although of course it is unrelated to any lunar cycle. But in Thailand an attempt has been made to synchronize women's cycles to the phase of the moon, using moonlight as a timely and romantic reminder of when to start on another course of 'the Pill'.

Comparisons of oestrous and menstrual cycles are made difficult because the external signs are reflections of different internal events. At one time, menstruation was thought to be a type of oestrus, and not until the 1930s did we begin to understand the true sequence of hormone changes during the cycle. Menstruation lasts about 4 or 5 days, and marks the end of the functional life of the corpus luteum; the first day of menstruation is regarded as day 1 of the new cycle (see Fig. 3-12). The follicular or proliferative phase follows the onset of menstruation, and although it normally lasts about 2 weeks, it can be extremely variable. Severe emotional stress can inhibit ovulation for weeks, months, or even years in some cases, and there is also a possibility that ovulation may be induced prematurely by the act of sexual intercourse. The moral of this story is that the 'rhythm' method of contraception is rather like Russian roulette.

If we look in more detail at the hormone changes during the menstrual cycle (see Fig. 3-13), we find that at the beginning of the follicular phase, the levels of FSH and LH are rising. As a result of this, a Graafian follicle begins to develop in one of the ovaries and the theca interna cells secrete an increasing amount of oestradiol-17β, androstenedione and 17α-hydroxy-progesterone. This latter steroid has only minimal biological activity, being an intermediate in the conversion of progesterone into androgens and oestrogens (see Chapter 1). The rising oestrogen secretion before ovulation begins to depress the secretion of FSH, and at the same time it causes the uterine glands to proliferate. Eventually, the oestrogen triggers off an ovulatory discharge of LH from the pituitary, usually accompanied by a simultaneous discharge of FSH. As in sheep, oestrogen secre-

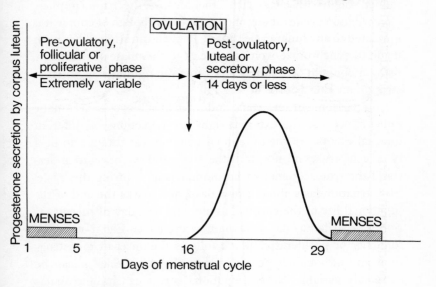

Fig. 3-12. The human menstrual cycle.

tion has declined by the time of the LH surge, and there is no marked pre-ovulatory secretion of progesterone.

We do not know the exact time intervals between the oestrogen peak, the LH peak and the moment of ovulation in women but they are probably similar to those in sheep. After ovulation has occurred, a corpus luteum begins to form and secrete large amounts of progesterone. It also secretes some 17α-hydroxy-progesterone and oestradiol-17β, so that the levels of all three steroids are elevated in the luteal or secretory phase of the cycle.

65

Role of hormones in sex cycles

Progesterone, or its metabolite, pregnanediol, is responsible for the post-ovulatory elevation of the basal body temperature, and oestrogen and progesterone together cause hypertrophy of the

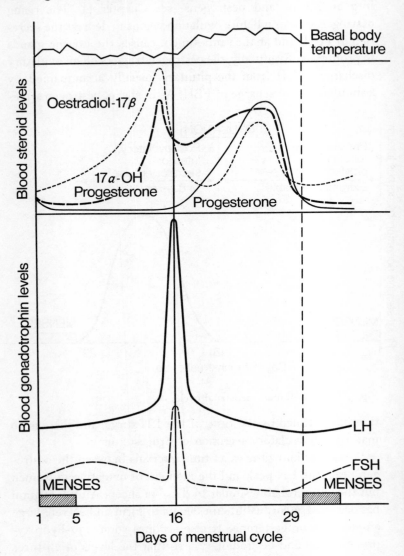

Fig. 3-13. Hormone changes in the human menstrual cycle.

endometrium and secretory activity of the endometrial glands in preparation for implantation of the egg, should fertilization occur.

The life of the corpus luteum was formerly thought to be fixed at about 14 days, but now that we can pinpoint the day of ovulation with blood LH assays, there are indications that some women can have short luteal phases of 8 days or less, when the corpus luteum secretes sub-normal amounts of progesterone. We still do not have a clear understanding of which pituitary gonadotrophins are necessary for the maintenance of the human corpus luteum; LH is certainly one of them, since it can stimulate progesterone secretion and prolong luteal life. But other factors must also be involved, because continuous treatment with LH or human chorionic gonadotrophin (HCG) can only prolong luteal life by a few days, and the corpus luteum will still regress during the course of the treatment. This is what happens at the end of a normal menstrual cycle, when there is no significant change in the blood levels of FSH or LH during luteal regression. Time will tell whether the newly discovered human prolactin is another key component of the pituitary luteotrophic stimulus.

In sheep, we saw how the life of the corpus luteum was ended by the luteolytic activity of uterine prostaglandin, even though the pituitary continued to secrete adequate amounts of luteotrophic hormones. Unfortunately, the same argument cannot be used in women. Whilst prostaglandin $F_{2\alpha}$ is indeed present in the human uterus at the end of the menstrual cycle, there is no evidence that prostaglandin causes the normal regression of the corpus luteum; for example, if the whole of the uterus is removed from a woman at hysterectomy, hormone assays suggest that she continues to have normal ovulatory cycles, and that the luteal phase is not prolonged.

Once the corpus luteum has started to regress, the hypertrophied secretory endometrium is suddenly deprived of its hormonal support; the endometrial arteries go into spasm, and the epithelium is sloughed off into the uterine lumen, together with a certain amount of blood. The prostaglandin in menstrual fluid probably has a stimulatory action on the smooth muscle of the uterus, which may account for the pains women experience at this time. Another consequence of luteal regression is that the hypothalamus escapes from the inhibitory effects of progesterone and oestrogen, so that the pituitary begins to secrete more FSH and LH in preparation for the next cycle.

There are still many mysteries to be unravelled in the human menstrual cycle. Some women are thought to have anovular cycles, when a corpus luteum never forms at all, and yet somehow they continue to menstruate normally at the end of each month. Today, the menstrual cycle of millions of women is controlled by 'the Pill' rather than the pituitary, and its mechanism of action is discussed by Malcolm Potts in Book 5, Chapter 2. The more we understand about the physiology of the normal menstrual cycle, the more likely we are to be able to modify it and so develop new approaches to contraception.

THE OESTROUS CYCLE OF THE RAT

The laboratory rat has been the work-horse of the experimental endocrinologist for decades, and all our ideas about the oestrous cycle were really derived from it. But in many respects, it is not a very good model to study. Although the rat is a spontaneous ovulator, like women and sheep, it differs from them by having extremely short oestrous cycles of only 4 days in length. This is because the corpora lutea that are formed following ovulation never really become functional; but if a mating takes place at oestrus, this stimulates the development of functional corpora lutea, and the animal either becomes pregnant or pseudopregnant, depending on whether or not the mating was fertile. Brian Heap discusses the question of pseudopregnancy in the next chapter. The rat also differs from sheep and women in that the whole timing of events in the oestrous cycle is determined by the diurnal rhythm; ovulation for example always occurs soon after midnight.

If we look at the hormone changes during the cycle (see Fig. 3-14), certain broad similarities are nevertheless apparent with the situation we have already encountered in sheep and women. A day before the rat is due to come into oestrus, on the morning of pro-oestrus, the secretion of oestradiol-17β from the developing Graafian follicle reaches peak values. This in turn triggers the ovulatory surge of LH on the afternoon of pro-oestrus, and ovulation occurs in the early hours of the following

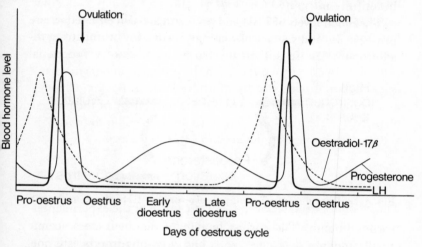

Fig. 3-14. Hormone changes in the oestrous cycle of the rat.

morning. The causal relationship of this sequence of events has been neatly demonstrated by giving rats an injection of anti-serum to oestradiol-17β during pro-oestrus, and showing that this abolishes the LH peak and hence prevents ovulation.

One of the most striking features of the rat's oestrous cycle is the enormous pre-ovulatory surge of progestin secretion (progesterone + 20α-dihydroprogesterone) that occurs. This is almost coincident with the LH peak, when the progesterone levels are far higher than during the short luteal (dioestrous) phase. Pre-ovulatory progesterone performs a most essential function in the rat; if the animal is to show oestrous behaviour, it must be under the influence of a high level of oestrogen, immediately

69

followed by a high level of progesterone (exactly the opposite of the situation in sheep). Since this progesterone must be secreted in pro-oestrus, after the old corpus luteum has regressed but before a new one has formed, the rat has made use of its 'permanent corpus luteum', the interstitial tissue. This can secrete large amounts of progesterone and 20α-dihydroprogesterone in response to LH stimulation, and since the release of LH is also facilitated by progesterone, the whole sequence of events is beautifully integrated (see Fig. 3-15).

The blood levels of FSH and prolactin are also elevated in late pro-oestrus, and the act of mating presumably stimulates gonadotrophin release still further so as to produce a functional

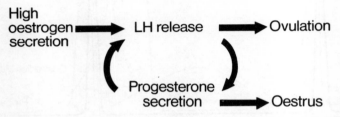

Fig. 3-15. Stimulatory effect of progesterone on LH release in the rat.

corpus luteum. The corpus luteum of dioestrus does secrete small amounts of progesterone and 20α-dihydroprogesterone, but we do not know what makes it secrete, or why it regresses after only 2 days, or indeed whether it serves any function at all.

So the story we have been trying to tell is essentially one of the declining influence of the corpus luteum. In sheep, it is the chief arbiter of cycle length; most of the oestrous cycle is occupied by the luteal phase, and a new ovulation occurs within about 3 days of luteal regression. In women, the luteal phase only occupies half the cycle, then there is a long follicular phase that can be disturbed by a variety of environmental factors. In the rat, the luteal phase has paled into insignificance, and there is a far greater degree of environmental control of the cycle, diurnal rhythms and sexual activity both having pronounced effects.

Human society has conditioned us into regarding menstrual

cycles as the norm, and pregnancy as an occasional untoward event, but we should remember that in any natural community of wild animals, pregnancy is the norm. The oestrous cycle is in a sense a pathological event that marks a failed bid for pregnancy, and is Nature's way of ensuring that another bid is made as soon as possible.

How is it that the male and the female can share the same pituitary and gonadal hormones, but show such clear differences in their patterns of secretion? The structure of the gonad undoubtedly plays a major role in determining the type of steroid that is secreted. Because of the transitory nature of the Graafian follicle and the corpus luteum, oestrogen and progesterone secretion waxes and wanes during the course of the oestrous or menstrual cycle, whereas testicular structure and activity are more constant. The sex differences in pituitary gonadotrophin secretion seem to result from the different responses of the male and female hypothalamus to gonadal steroids; without this feedback, the hypothalamus is incapable of meaningful, coordinated control of gonadotrophin release. Many of the perplexing species differences in the reproductive patterns of the female can be traced back to the love–hate relationship that exists between the hypothalamus, the corpus luteum and the uterus. A corpus luteum is essential for the initiation of a pregnancy, but is an embarrassment if the bid for pregnancy is unsuccessful, and different groups of mammals have tackled this problem in different ways.

There is still much that we do not understand, such as the role of oestrogen and prolactin in the male, or androgen in the female. But we can expect rapid advances in the next few years now that we have a new generation of simple, sensitive and specific methods of hormone assay.

SUGGESTED FURTHER READING

Biochemistry of semen and of the male reproductive tract. T. Mann. London; Methuen (1964.)

71

Role of hormones in sex cycles

Reproduction. R. V. Short. *Annual Reviews of Physiology* **29**, 373 (1967).

Pituitary and gonadal hormones in women during spontaneous and induced ovulatory cycles. G. T. Ross, C. M. Cargille, M. B. Lipsett, P. L. Rayford, J. R. Marshall, C. A. Strott and D. Rodbard. *Recent Progress in Hormone Research* **26**, 1–48 (1970).

Mechanisms regulating the menstrual cycle in women. R. L. Vande Wiele, J. Bogumil, I. Dyrenfurth, M. Ferin, R. Jewelewicz, M. Warren, T. Rizkallah and G. Mikhail. *Recent Progress in Hormone Research* **26**, 63–95. (1970).

Factors affecting the secretion of steroids from the transplanted ovary of the sheep. J. A. McCracken, D. T. Baird and J. R. Goding. *Recent Progress in Hormone Research* **27**, 537 (1971).

Relationship between blood levels of luteinising hormone and testosterone in bulls, and the effects of sexual stimulation. C. B. Katongole, F. Naftolin and R. V. Short. *Journal of Endocrinology* **50**, 457 (1971).

4 Role of hormones in pregnancy
R. B. Heap

Hormones have a major part to play in the maintenance of pregnancy. In this chapter we shall be concerned with how hormones regulate the maternal adjustments to pregnancy, and how their secretion is controlled to meet the changing needs. Aspects of fertilization, blastocyst nutrition, implantation and lactation will be mentioned only briefly since they are discussed in other chapters.

The role of hormones in pregnancy became prominent with the evolution of viviparity, when a series of innovations first appeared: reduction in the yolk content of eggs, formation of the placenta, retention of the young within the female genital tract, and parental care of offspring. Endocrine control resides primarily in the hormones of the pituitary, the ovary and the placenta, and a distinction can be drawn between these organs and those, such as the thyroid, adrenal and parathyroid whose role is more of a supporting or permissive one.

MATERNAL RECOGNITION OF PREGNANCY

The presence of a corpus luteum actively secreting progesterone is a characteristic feature of early pregnancy. In marsupials, such as the opossum and red kangaroo, the life-span of the corpus luteum in the pregnant and non-pregnant animal is similar, and the gestation period is about the same length as the oestrous cycle. In eutherian mammals, however, a requirement of pregnancy is that the luteal phase of the oestrous cycle should be prolonged and ovulation suppressed. However, the important question of how an animal knows that it is pregnant, and how it prolongs its oestrous cycle and delays the recurrence of ovulation, cannot yet be satisfactorily answered.

73

Role of hormones in pregnancy

Evidence on the maternal recognition of pregnancy is available in many species. In the mare, only fertilized eggs are allowed to pass into the uterus, which suggests an extremely early recognition of pregnancy. In other species, some signal is given by the blastocysts after they have entered the uterus, but before they have implanted. In sheep, the message is transmitted 2 or 3 days before the animal would normally return to oestrus; the conceptus appears to neutralize the potent luteolytic factor manufactured by the uterus, which otherwise would cause the corpus luteum to regress. In women and rhesus monkeys, pregnancy is probably recognized by the luteotrophic activity of the conceptus, which prolongs the life of the corpus luteum and delays menstruation. There are some species of eutherian mammals like the ferret and dog that emulate the marsupials by having a luteal phase of similar length to the gestation period, so that 'pregnancy recognition' may not even be necessary.

Some species show a phenomenon known as delayed implantation (Book 2, Chapter 1), where the period of pre-implantation may exceed that of post-implantation. The delay may be caused by lactation, as in the rat, mouse and kangaroo (facultative delay), or by environmental changes, as in the badger or roe deer (obligatory delay). The delay found in lactating rats is related to the number of young suckled and does not usually occur if the litter size is less than five; implantation can be induced by a single injection of a minute amount of oestrogen. The termination of delayed implantation in the badger is associated with an increased secretory activity of the corpus luteum, but in the roe deer there is no evidence to suggest that changes in either oestrogen or progesterone secretion are responsible for ending the delay.

Faced with such a baffling array of species differences, we are forced to conclude that there is not a single mechanism for the recognition of pregnancy that is common to all species.

ENDOCRINE ORGANS IN PREGNANCY

The ovary and the corpus luteum

The corpus luteum is a characteristic, though not a unique,

74

feature of mammalian pregnancy. Luteal bodies have been described throughout the vertebrates, including elasmobranch and teleost fishes, but they do not always function as endocrine glands of gestation. The mammalian corpus luteum plays a specific role in the endocrinology of gestation through the synthesis and secretion of progesterone, 'the hormone of pregnancy'.

The first indication of the role of the corpus luteum in pregnancy maintenance came at the beginning of this century when Ludwig Fränkel demonstrated that removal of the corpora lutea from a pregnant rabbit terminated gestation. He was fortunate to choose the rabbit for in this species ovariectomy always causes abortion; in others, such as the guinea pig, sheep or horse, pregnancy may continue if the operation is performed after a certain time in gestation (Table 4-1). But there is no known species in which pregnancy can be maintained without progesterone, except perhaps the elephant (Book 4, Chapter 1).

In species that usually give birth to a single offspring (monotocous species), such as man and cow, the number of corpora lutea usually corresponds to the number of embryos, but in the mare the corpus luteum of pregnancy is supplemented by several accessory corpora lutea, formed early in gestation either by ovulation or by luteinization of unruptured follicles. In highly polytocous species, such as rabbit and pig, some embryonic mortality is commonly found and the number of corpora lutea exceeds the number of embryos. An extreme example of such a discrepancy is found in the plains viscacha, a hystricomorph rodent from S. America related to the guinea pig, where several hundred follicles ovulate at a time, giving rise to a mass of lutein tissue, although only one or two blastocysts normally implant and develop.

Regulation of progesterone secretion

Progesterone is the most active of the naturally occurring progestagens and it is produced by the ovary of most species in greater quantities than any other steroid. An exception is to be

TABLE 4-1. Removal of the ovaries or pituitary in pregnancy and its effect on the maintenance of gestation in various species. +, Fetuses survive; ±, some fetuses survive; −, abortion.

Animal and length of gestation (days)	Approx. stage of pregnancy when operation performed			
	Ovariectomy		Hypophysectomy	
	First half	Second half	First half	Second half
Sheep (148)	−	+	−	+
Woman (280)	+	+	+	+
Monkey (165)	+	+	+	+
Guinea pig (68)	±	+	+	+
Rat (22)	−	±	±	+
Mouse (23)	−	−	−	+
Hamster (16)	−	−	−	+
Cat (63)	−	±	n.d.	±
Cow (282)	−	±	n.d.	n.d.
Ferret (42)	−	±	−	±
Horse (350)	−	+	n.d.	n.d.
Dog (61)	−	n.d.	−	±
Goat (148)	−	−	−	−
Sow (113)	−	−	−	−
Rabbit (28)	−	−	−	−
Opossum (16)	−	−	n.d.	n.d.
Armadillo (150)	Implantation may occur	−	n.d.	n.d.

n.d. = not determined

found in the pregnant rabbit where the ovarian secretion rate of a reduced metabolite, 20α-dihydroprogesterone, may be greater than that of progesterone. This metabolite is only a weak progestagen and is secreted by the interstitial tissue of the rabbit ovary. In the pregnant rat, where the secretion rate of 20α-dihydroprogesterone is about half that of progesterone, it is synthesized principally by corpora lutea. Apart from the hormone secretions of the corpus luteum and interstitial tissue, a third steroid-secreting tissue in the ovary, the theca interna of unruptured follicles, undergoes waves of activity during gestation, presumably associated with the secretion of ovarian oestrogens.

The corpus luteum of pregnancy is composed of morphologically distinct cells, large and rounded in appearance and well endowed with a pale-staining cytoplasm. The ultrastructure of the actively secreting luteal cell resembles that of all steroid-secreting cells: the cytoplasm is packed with smooth endoplasmic reticulum and contains lipid droplets and an enlarged Golgi complex. The lipid droplets probably represent intracellular stores of steroid precursors, depleted during maximum synthesis, as in pregnancy, and restored at times of reduced steroid production, as during luteal regression. During peak activity, luteal cells in the horse and guinea pig may secrete about 300 pg $(3 \times 10^{-10}g)$ progesterone/cell/day, or 7×10^{14} molecules/cell/sec! The tissue concentration of progesterone closely reflects secretory activity in most species. It is normally low (about 30 $\mu g/g$), though it may differ considerably between species; in the common zebra the concentrations are higher than average, whereas in the elephant they are appreciably lower. Whether the very low level in the elephant is because of an economical requirement for progesterone in pregnancy, or because of the existence of an unidentified progestagen, is a tantalizing question that waits to be answered. The wide range of progesterone concentrations in luteal tissue suggests that transport down a diffusion gradient into the blood stream is not the only mechanism involved in the release of progesterone

Role of hormones in pregnancy

from its site of synthesis. For example, intracellular binding to tissue proteins may also play some part in regulating the rate of secretion. In many species, however, the total amount of progesterone stored in the corpus luteum is small and is equivalent to only a few minutes' supply, and this is true of steroidogenic tissues in general.

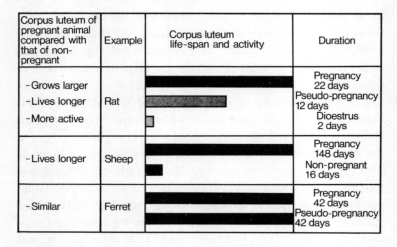

Corpus luteum of pregnant animal compared with that of non-pregnant	Example	Corpus luteum life-span and activity	Duration
- Grows larger - Lives longer - More active	Rat		Pregnancy 22 days Pseudo-pregnancy 12 days Dioestrus 2 days
- Lives longer	Sheep		Pregnancy 148 days Non-pregnant 16 days
- Similar	Ferret		Pregnancy 42 days Pseudo-pregnancy 42 days

Fig. 4-1. Three variants of the growth and function of the corpus luteum in pregnancy.

Nature of the luteotrophic complex

We must consider as distinct processes the formation and survival of the corpus luteum as a structure, its synthetic abilities and its secretory activity in terms of progesterone and other metabolites. Fig. 4-1 illustrates three variants of the growth and function of the corpus luteum encountered in pregnancy. It may grow larger, survive longer and be more active than during the normal cycle or pseudopregnancy (rat and guinea pig); or neither its size nor its activity may increase, although its life-span is prolonged (sheep, cow and pig); or its life-span, growth and function may be indistinguishable from a corpus luteum of pseudopregnancy (ferret, dog).

78

The prolongation of luteal life-span and function in pregnancy is largely regulated by the secretion of a luteotrophic hormone complex. In some species (rabbits, ferrets, pigs and goats) this complex of protein hormones is derived principally from the pituitary, since hypophysectomy always results in abortion (Table 4-1). In other animals (sheep, rats and guinea pigs) the pituitary is only necessary at the beginning of pregnancy, since placental luteotrophins are able to compensate for its removal later on.

The early demonstration in hypophysectomized pregnant rats that injections of prolactin could maintain functional corpora lutea and prevent abortion led to the view that prolactin was the luteotrophic hormone. More recently it has become apparent that in most species, the luteotrophic stimulus is made up of a hormone complex in which prolactin plays an important part. A consideration of the nature of this luteotrophic complex in five species will illustrate the diversity of the endocrine mechanisms that have developed to maintain the ovarian secretion of progesterone and thereby ensure the maintenance of gestation.

Rabbit. The corpora lutea soon degenerate after hypophysectomy, and prolactin plus FSH and possibly low levels of LH are necessary to support luteal function (Fig. 4-2a). One of the roles of this luteotrophic complex is to stimulate the follicles to secrete oestrogens, which in turn have a direct trophic influence on luteal cells, prolonging their life and promoting progesterone secretion. When the secretion rate of progesterone declines at the end of gestation, pituitary LH production is raised. This increases the secretion of 20α-dihydroprogesterone by the interstitial tissue, and LH may also hasten luteal regression by the destruction of oestrogen-secreting follicles. Anything that affects the secretion of the luteotrophic complex (hypophysectomy or stalk section) or stops ovarian oestrogen secretion (destruction of ovarian follicles by X-irradiation) will terminate pregnancy.

Role of hormones in pregnancy

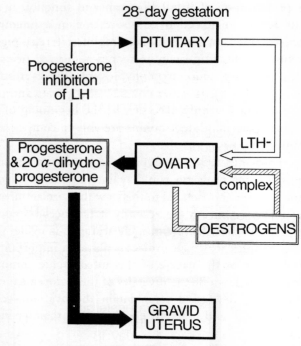

Fig. 4-2a. The hormonal maintenance of gestation in the rabbit, showing the origin of the luteotrophic complex that maintains the corpus luteum, and of the progesterone that maintains the pregnant state. Arrows indicate the relative importance of the different pathways.

Rat. Prolactin and LH are the chief components of the pituitary luteotrophic complex during the first half of gestation (Fig. 4-2b). They are supplemented in the second half of gestation by a prolactin-like substance derived from the placenta. If the corpus luteum of pregnancy is to maintain a relatively high secretion rate of progesterone, 20α-hydroxysteroid dehydrogenase, a soluble enzyme in the corpus luteum responsible for the conversion of progesterone to 20α-dihydroprogesterone, must be suppressed. Prolactin, which inhibits this enzyme, and LH, which stimulates steroidogenesis, thus regulate the activity of the corpus luteum. Towards the end of pregnancy, the withdrawal of placental luteotrophins, or the increased release of pituitary LH, leads to a fall in ovarian progesterone secretion and to a rise in secretion of 20α-dihydroprogesterone just before the onset of parturition. This steroid switch is also important for initiating lactation.

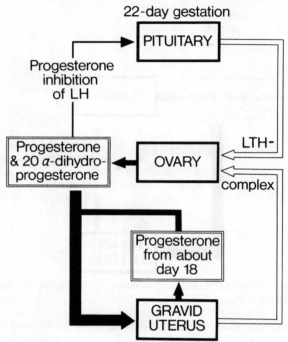

Fig. 4-2*b*. The hormonal maintenance of gestation in the rat (compare with Fig. 4-2*a*).

Sheep. In this species the conceptus prolongs the life of the corpus luteum by neutralizing the uterine luteolytic factor. Up to about day 50 of pregnancy the conceptus is unable to produce sufficient luteotrophins to maintain pregnancy, since hypophysectomy before that time causes abortion (Fig. 4-2*c*, Table 4-1). Both prolactin and LH appear to be necessary for luteal function.

Pig. In this species, the nature of the luteotrophic complex changes during gestation. Early in pregnancy, the corpora lutea

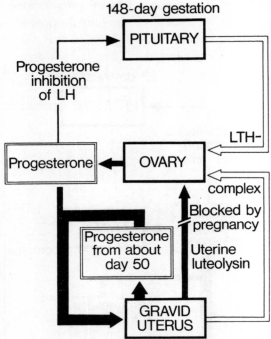

148-day gestation

PITUITARY

Progesterone
inhibition
of LH

LTH–

Progesterone ◄— OVARY

complex

Blocked by
pregnancy

Progesterone
from about
day 50

Uterine
luteolysin

GRAVID
UTERUS

Fig. 4-2c. The hormonal maintenance of gestation in the sheep (compare with Fig. 4-2a).

are relatively independent of pituitary control and their survival depends on the ability of the conceptus to neutralize the uterine lytic factor. Later, up to about day 40, LH is the predominant luteotrophic factor, but thereafter prolactin becomes increasingly important. Oestrogens also may play a part in the maintenance of the corpus luteum, possibly acting indirectly to stimulate pituitary prolactin secretion; the placenta of the pig produces appreciable amounts of oestrogen in late pregnancy.

Woman. In women, the formation of the corpus luteum of pregnancy probably depends upon the secretion of human chorionic gonadotrophin (HCG) by the trophoblast within a few days of implantation, together with human chorionic somatomammotrophin (HCS) (Fig. 4-2d) (see Chapter 1).

82

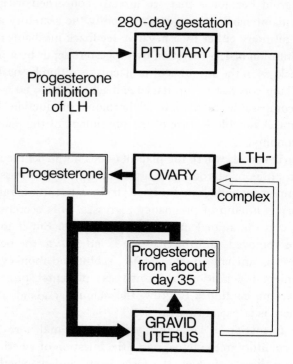

280-day gestation

PITUITARY

Progesterone
inhibition
of LH

Progesterone

OVARY

LTH–
complex

Progesterone
from about
day 35

GRAVID
UTERUS

Fig. 4-2*d*. The hormonal maintenance of gestation in the woman (compare with Fig. 4-2*a*.)

Placental hormone production

We have just seen how the placenta in some species may assume some of the functions of the pituitary, and so control the corpus luteum. But in addition, it may assume some of the endocrine functions of the ovary. This is best seen in species where ovariectomy does not interfere with pregnancy (Table 4-1). In women and monkeys, the placenta produces enough progesterone to maintain pregnancy, at a very early stage, so that the role of the corpus luteum in pregnancy maintenance is shortlived.

The regulation of gestation comprises a series of complex interactions between protein hormones that are luteotrophic,

83

and steroid hormones that are directly concerned with pregnancy maintenance. Steroids also modify the secretory activity
of the pituitary either by a negative feedback mechanism, as in
the inhibition of LH secretion by progesterone, or by a positive
feedback, as in the stimulus of prolactin secretion by oestrogen.
In striking contrast to pituitary feedback, we have no evidence
of a comparable regulation of hormone production by the
placenta. A notable feature of the physiology of the placenta is
its autonomy.

Even in species where the placenta does not make recognized
luteotrophins or steroids (e.g. goat, rabbit), it still possesses a
very important endocrine function; if the conceptus is removed,
the corpus luteum of pregnancy regresses, milk secretion may
begin, and the animal may return to oestrus. But if the fetus
alone is removed, and the placenta is left *in situ*, the period of
gestation is unchanged (rat, mouse, rabbit and monkey). This
experimental condition is known as 'placental pregnancy';
the placenta continues to grow, the mammary glands develop
and there is a 'pseudo-parturition'.

The formation of large quantities of placental hormones in
close proximity to the fetus, raises the question of whether they
have any direct effect on the fetus itself, and of whether the
fetus is protected against the high levels of these potent substances. The amounts of protein and steroid hormones stored in
the placenta are generally low and are equivalent to only a few
minutes' supply. In man, there is an effective intra-placental
barrier to the transfer of protein hormones, such as HCS,
though small quantities of HCG may reach the fetus. In contrast
there is an appreciable transfer of progesterone (Fig. 4-3) and,
to a lesser extent, of oestrogens to the fetus. These steroids are
rapidly metabolized in the fetus, principally in the adrenals,
liver and peripheral tissues, by hydroxylation or conjugation,
mechanisms that effectively reduce the biological potency of the
parent compound. In woman, the transfer of progesterone from
the placenta and of glucocorticoids from the mother into the
fetal circulation may have an important physiological role in

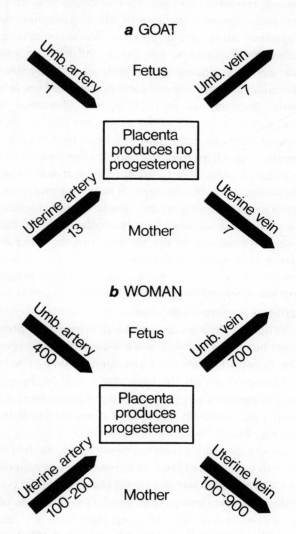

Fig. 4-3. Intriguing contrasts in the placental production of progesterone in two species. Typical concentrations of plasma progesterone (ng/ml) in the uterine artery and vein, and in the umbilical artery and vein.

safeguarding the fetus from adrenal insufficiency, since its own adrenals suffer from a deficiency of Δ^5-3β-hydroxysteroid dehydrogenase, an essential enzyme in the biosynthesis of corticosteroids. In human pregnancy the rate of placental hormone secretion appears to be correlated with the size of the placenta, as is evident from the relation of the blood levels of total oestrogens, progesterone and HCS, and placental weight (Fig. 4-6).

Intriguing contrasts exist between species in the ability of the placenta to secrete steroid hormones (Fig. 4-3). In some instances the placenta fails to produce progesterone, though it forms large quantities of oestrogens throughout gestation (as in the pig). The sheep and goat are closely related to each other, and the placenta is morphologically similar in both, but progesterone is synthesized in the placenta of the former but not of the latter (Fig. 4-2*c* and Fig. 4-3).

Pituitary and other endocrine organs

The pituitary has an indispensable role in the secretion of hormones that are concerned with the early events of pregnancy but, as we have seen, its role is transferred in some species to the placenta in the later stages of gestation (Table 4-1). Relatively little is known about the changes that occur in the pituitary content of hormones during pregnancy and recent work has been concerned more with the changes in hormone concentrations in circulating blood.

Pregnancy is associated with an increased size of the pituitary in some species. The cytology of the pituitary changes during gestation and the appearance of specialized 'pregnancy-cells' is one of the most characteristic findings. The function of these cells is uncertain but may be concerned with an increased secretion of a pituitary luteotrophic complex. The cells are believed to be derived from the chief, or 'chromophobe', cells, but the difficulty of obtaining a method that will specifically localize protein hormones such as prolactin precludes an accurate assessment of their role.

The function of the other endocrine organs (thyroid, para-thyroid, pancreas, adrenals) is related to the differing conditions imposed by pregnancy rather than to the fact that they have a major role in the maintenance of gestation. The adrenals, frequently heavier in females than in males, may become more active and enlarge. Biochemically, they have the capacity to produce appreciable amounts of gonadal steroids (oestrogens, androgens and progestagens). The concentration of these steroids in adrenal venous blood is high under the conditions of stress induced experimentally by surgery, anaesthesia or ACTH administration, but whether they are high in the normal, unstressed condition is in doubt. If gonadal hormones of adrenal origin played a significant part in pregnancy maintenance, then adrenalectomy would be expected to result in the termination of gestation. On the contrary, experiments first performed in the dog show that adrenalectomy is tolerated better when the animal is pregnant than at other times. This is probably because the secretory activity of the corpus luteum of pregnancy compensates in some way for the loss of corticosteroids. Similar results have been obtained in non-pregnant animals, which also survive longer after adrenalectomy if treated with progesterone.

HORMONES AND THE UTERUS IN PREGNANCY

Growth of the uterus

Though progesterone is frequently referred to as the 'hormone of pregnancy', it has also been called 'the useless hormone' because it rarely acts alone. Synergistic interactions are usually encountered when the concentrations of progesterone and oestrogens differ by about three orders in favour of progesterone. As the difference in concentrations decreases, their actions become increasingly antagonistic. The histological changes that are observed in target organs during gestation are the result of interactions between oestrogens and progesterone. In the

pregnant rabbit there is extensive proliferation of the endometrium due to the 'priming' effect of oestrogens and the
subsequent action of progesterone on the luminal and glandular
epithelium. This proliferation prepares the endometrium for
the reception of the embryo and is associated with the events of
implantation. Endometrial proliferation is a prominent feature
of the progestational state in the cat, ferret and rabbit, but in
other species the degree of proliferation is much less pronounced
(rat, mouse, guinea pig, sheep and cow). It can be reproduced
experimentally in ovariectomized rabbits by treatment with
appropriate doses of oestrogen followed by progesterone, though
large doses of progesterone alone will sometimes produce
similar effects. This synergistic interaction between oestrogen
and progesterone causes distinctive changes in the histological
appearance of other target organs during pregnancy, notably
the mucification of the vagina and the growth of alveoli and
ducts in the mammary glands.

The growth of the uterus during gestation is principally an
enlargement in the size of the myometrium. This results from
growth of existing cells, rather than from increase in cell numbers. So far as the uterus is concerned, we understand best the
action of oestrogens on its growth and composition. If ovariectomized rabbits are injected with oestradiol-17β, protein synthesis
is stimulated and the RNA : DNA ratio of the endometrium
increases, as in normal pregnancy. If progesterone is injected,
there is no effect on uterine growth, whereas treatment with a
suitable combination of both hormones may promote even larger
increases than are found with oestrogen alone. This influence of
oestrogens on the growth and weight of the uterus is rapid; within 30 minutes of injection into rats, uterine hyperaemia occurs,
and within 48 hours, dry weight, protein synthesis and glycogen
deposition increase progessively. This 'uterotrophic effect' is a
direct one and is not affected by hypophysectomy or
adrenalectomy.

Uterine growth in pregnancy cannot be entirely accounted
for by the effects of oestrogen and progesterone. Whereas the

contractile protein actomyosin, found in the myometrium, increases in concentration under the influence of oestrogen, a major stimulus for uterine growth during gestation is believed to consist of the 'stretch' induced by the growing fetus and its associated membranes.

Inhibition of myometrial activity

Progesterone has a multiple role in the maintenance of gestation. It delays ovulation by the suppression of LH secretion; it acts on the endometrium to prepare for the reception of the embryo; and it renders the contractile myometrium quiescent so that implantation proceeds normally and the expulsion of embryos is prevented. Arpad Csapo has called the latter property the 'progesterone-block' of myometrial activity.

Experiments in rabbits first showed that when the uterus is dominated by the influence of progesterone, as in pregnancy or when the steroid is applied locally to the uterus *in vivo* or *in vitro*, coordinated myometrial contraction is reduced. Moreover, the stimulatory effect of oxytocin on uterine contractility is abolished. This effect is seen in experiments where oxytocin is given to a rabbit on the 30th day of gestation. The treatment fails to induce parturition, but when repeated a day later, it provokes a normal delivery. On the other hand, if progesterone is given 24 hours before the oxytocin injection, the onset of parturition is delayed. In other studies, the increase in myometrial activity associated with the onset of normal parturition was delayed in a unilateral fashion if progesterone was infused locally into one gravid horn. Yet another investigation suggested that progesterone acted not only locally to reduce coordinated myometrial contractility and electrical activity, but that it also inhibited the hypothalamic release of oxytocin.

Whether the 'progesterone-block' hypothesis applies equally in species other than the rabbit is still unclear. Effects of progesterone on myometrial contractility have been reported in the rat, sheep and man, though they are not always the same as those found in the rabbit. The guinea-pig is exceptional in that progesterone has little effect on myometrial activity even when it is applied locally in very high doses.

Nevertheless the role of progesterone in reducing myometrial activity in the rabbit is convincing, and widely differing theories have been put forward to explain how this happens. Csapo has proposed that progesterone 'hyperpolarizes' the myometrial cell, reducing its excitability. Consequently a decline in progesterone, as at term, would facilitate myometrial excitation. Changes in the ionic permeability of the myometrial cell induced by progesterone are more likely to underlie the inhibition of activity.

Progesterone levels before parturition

If one of the functions of progesterone in pregnancy is to block myometrial activity, the increased contractility found at parturition should be associated with a removal of this inhibition, and the problem is discussed in some detail in Book 2, Chapter 3. Some observations do not support the view that parturition is initiated by a sudden fall in plasma progesterone levels in all species; for example, there is no such fall in women, and progesterone injections cannot prevent parturition in guinea pigs. The interactions of other factors such as the binding of progesterone to tissue and plasma proteins, and the metabolism of progesterone in the endometrium and myometrium, probably also play a part in the normal processes concerned with the maintenance of gestation. The activity of the fetal adrenal glands may lead to events that induce prostaglandin release and trigger parturition in a number of species.

Stimulation of myometrial activity

Oestrogen stimulates the contractile activity of the quiescent, immature uterus, causing regular, rhythmic contractions. At, or just before, the time of parturition this activity becomes synchronized, and with the release of oxytocin the contractions increase in both frequency and amplitude (see Book 2, Chapter 3).

The increase in myometrial activity in late pregnancy corresponds with the time when the concentration of oestrogens in blood reaches its highest levels (rat, guinea pig, sheep, goat and human female). In the sheep there is a very sharp rise in the

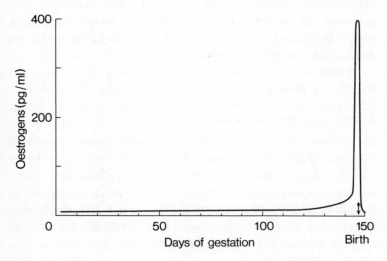

Fig. 4-4. A sharp rise in the plasma concentration of total, unconjugated oestrogens in the pregnant sheep just before parturition. The arrow indicates the time of parturition. Oestrogens were not detectable before the 50th day of gestation by the radioimmunoassay employed (< 5 pg/ml).

levels of total, unconjugated oestrogens about 48 hours before parturition (Fig. 4-4). This is also the time when spontaneous uterine activity and reactivity to oxytocin show a dramatic increase. It appears that the removal of the 'progesterone-block' of myometrial activity towards the end of gestation may be related not only to a reduction in the local concentration of progesterone, but also to the antagonism of a raised oestrogen level.

The administration of oestrogens during gestation causes

abortion in several species including the rat, mouse, rabbit, cat and cow. The abortifacient effect depends on the time of pregnancy when the treatment is started. Late pregnancy is much more difficult to interrupt in this way, presumably because of the greater concentrations required to overcome progesterone dominance. Oestrogens probably do not have a similar importance in the termination of pregnancy in all species; in some, they appear to be without effect on the duration of gestation, or they may even prolong it slightly. In pregnant women fetal death results in a drastic reduction of oestrogen synthesis, but labour can be induced without oestrogen therapy.

In summary, progesterone plays a dominant role in the maintenance of pregnancy in eutherian mammals. A number of endocrine changes are implicated in the mechanism by which normal pregnancy is terminated. There may be a reduction in the 'progesterone-block' of myometrial activity; an increase in oestrogen production towards the end of gestation, thereby affecting the synergistic balance of oestrogen and progesterone; release of oxytocin and the stimulation of myometrial activity; an activation of the fetal adrenohypophysial axis and an elevated secretion of fetal adrenocorticosteroids; prostaglandin release; and the relaxation of the pubic symphysis promoted by the action of relaxin.

Uterine blood flow

Uterine blood flow increases during gestation but when corrected for the increased weight of the uterus and its contents, the relative blood flow apparently decreases in mid-pregnancy to fairly stable levels which are maintained until parturition (e.g. sheep, goat). Uterine blood flow is probably influenced by hormonal factors, apart from the demands of the growing fetus on maternal nutritive supplies.

Investigations in sheep show that the uterine capillary bed is highly sensitive to oestrogens, which cause a striking decrease in vascular resistance and increase in blood flow. Autoregulation of uterine blood flow is lacking, but in the pregnant animal the low resistance of the placenta is thought to provide an adequate regulation of the response to vaso-active stimuli.

OTHER EFFECTS OF HORMONES IN PREGNANCY

Metabolic changes

Pregnancy is an anabolic state and is associated with an increased metabolic activity of the maternal organism. The characteristic gain in weight is due partly to the growth of the uterus and fetus, and partly to increases in blood volume and in the weights of organs such as the liver and mammary glands. In gestation the increased body retention of water, protein and fat is hormonally regulated, and results from the presence of the placenta rather than the fetus. Thus, maternal weight gain in pregnancy is not simply an accumulation of excess reserves to protect the mother against the ever-increasing demands of the fetus, but results from the physiological adjustments of pregnancy controlled by the changing hormonal conditions.

Progesterone is believed to play an important part in maternal weight gain during gestation, but the mechanism of its effect is obscure. The mouse and rat have been studied in greatest detail, but even in these two species there are different explanations of the way progesterone can influence body weight. In the mouse, progesterone induces water retention, stimulates appetite and food intake, and has a protein anabolic effect. In the rat, weight gain is only observed if ovarian function is suppressed either by the administration of progesterone or by ovariectomy. Yet another effect is observed in the human female, where progesterone is catabolic and may promote increased salt excretion. On the other hand recent reports suggest that some women taking oral progestagens gain weight. The metabolic effects of progesterone during pregnancy are complex; they may even differ according to the stage of gestation.

Role of hormones in pregnancy

HORMONE PRODUCTION IN PREGNANCY

The increased production of hormones in gestation probably reflects the changing requirements of the mother. The extent of these increases is indicated by the measurement of hormones in blood, or by the estimation of the daily excretion of hormones, or, preferably and more specifically, by the estimation of hormone production rates by tracer kinetic techniques. The amounts of hormones in blood reflect the secretory activity of endocrine organs more closely than urinary levels of hormone metabolites. What is more important, they reflect the concentration of hormones transported to target tissues such as the uterus and mammary gland. With the advent of the new and highly sensitive methods of radioimmunoassay and competitive protein binding, day-to-day changes in the blood levels of hormones can be monitored throughout gestation. Pregnancy may be diagnosed very early by the measurement of hormone levels; immunological tests for HCG are routinely used for pregnancy diagnosis in women, but in the larger domestic animals the identification of the conceptus by rectal palpation is commonly used in clinical practice and is probably the most reliable indication of pregnancy (Table 4-2).

Because of their high lipid solubilities, oestrogens and progesterone are transported in blood largely bound to plasma proteins. Some proteins, such as albumin, have a very large capacity for steroid binding but their affinity is low; others, such as transcortin (corticosteroid binding globulin, CBG), have a low capacity but a high affinity. In consequence, only a small proportion of the oestrogen and progesterone measured in blood is present in a form that is free or non-protein bound. The presence of plasma proteins in gestation with high affinities for steroid hormones affects the rate of metabolism of a steroid and its concentration in blood, and also its availability to target tissues. Therefore, we should consider the rates at which hormones are produced and metabolized in the body since this affects their role in gestation.

94

TABLE 4-2. Clinical methods of pregnancy diagnosis in different species.

	Test	Source of material	Substance tested	Test is applicable from the following week after fertilization
Woman	Immunological	Urine or plasma	HCG + LH	3rd
		Urine or plasma	HCG only	1st
Cow	Rectal palpation	—	—	6th
Ewe	Radiography	—	—	8th
	Ultrasonics	—	—	9th
Sow	Rectal palpation	—	Enlargement of middle uterine artery	4th
	Radiography	—	—	12th
	Ultrasonics	—	—	8th
	Histological test	Vaginal mucosa	Cell height	4th
Mare	Rectal palpation	—	—	3rd
	Immunological	Plasma	PMSG	6th
	Steroid assay	Urine	Oestrogens	20th

Role of hormones in pregnancy

Kinetics of hormone metabolism

Under steady-state conditions the amount of a hormone produced is equal to the rate at which it is destroyed. The rate of
destruction can be measured by several techniques, including
one in which a tracer quantity of isotopically labelled steroid is
infused at a constant rate for several hours. When the ratio
of the isotopic and endogenous hormone concentration in
blood is constant, the metabolic clearance rate (MCR) is given
by the equation,

$$\text{MCR (l/min)} = \frac{\text{Rate of infusion of labelled compound, } \mu c/min}{\text{Blood concentration of labelled compound, } \mu c/l}$$

Metabolic clearance rate is defined as the volume of blood
that is completely and irreversibly cleared of a compound in
unit time. The production rate of the hormone ($\mu g/min$) is
calculated by multiplying the metabolic clearance rate by the
blood concentration. An alternative way in which hormone
production may be measured consists of the estimation of the
urinary excretion of a metabolite, such as pregnanediol or
oestriol. Whereas both methods have their particular applica-
tions, the information derived from the urinary procedure is
less direct and may be complicated if the excreted metabolite
is derived from several precursors. Thus, urinary pregnanediol
in women may be formed from several precursors of which
progesterone is quantitatively the most important, though only
in pregnancy.

Progesterone. The concentrations of this steroid in blood during
gestation differ appreciably between various species. There is a
100-fold rise over the non-pregnant levels in women and guinea
pigs, a 10-fold increase in sheep, and no rise over pseudopregnant
levels in ferrets (Fig. 4-5).

There are at least two ways in which the progesterone
requirements of pregnancy may be met – either an increase in
progesterone production (women), or a reduction in its rate of
metabolism (guinea pigs). The sources of production are very

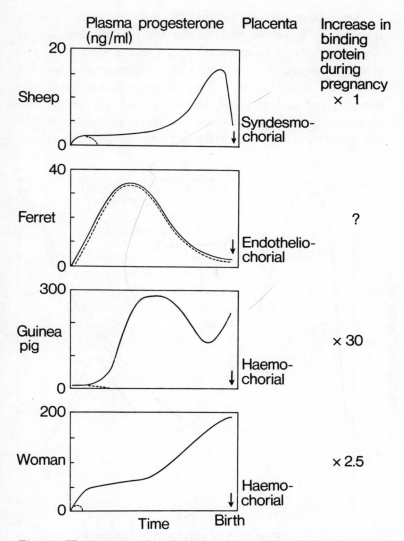

Fig. 4-5. The concentration of plasma progesterone in different species, pregnant (solid lines) and non-pregnant (dotted lines). The arrows indicate the time of parturition. The figures for the binding protein concentration (corticosteroid-binding globulin, CBG) represent the typical change observed in the gravid compared with the non-gravid condition.

Role of hormones in pregnancy

similar in these two species; initially, the corpus luteum is the main site of secretion, but the placenta becomes increasingly important later in gestation (Figs. 4-5 and 4-6). Yet the rate at which progesterone is metabolized differs greatly between them. In women, the MCR of plasma progesterone is similar in non-pregnant and pregnant subjects (about 1.8 l/min). The plasma progesterone concentrations increase gradually throughout gestation (Fig. 4-5), reflecting the increasing secretory

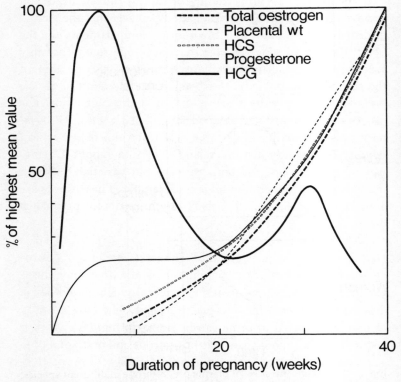

Fig. 4-6. Hormone concentrations in blood during gestation in the human female, and their relation to placental weight. Typical values of the highest mean concentrations reported: progesterone, 200 ng/ml; total oestrogens, 160 ng/ml (individual oestrogens mainly present in a conjugated form, oestrone, 40 ng/ml, oestradiol-17β, 10 ng/ml, oestriol, 100 ng/ml); HCG, 50 i.u./ml; HCS, 10 μg/ml; placental weight, 700 g.

activity of the placenta. In the guinea pig, the progesterone concentration rises sharply after the 15th day of gestation (Fig. 4-5). The rise is accompanied by a pronounced fall in MCR and raised concentrations of specific binding proteins in blood which have a high affinity for progesterone (e.g. progesterone binding protein, PBP). This 'progesterone-conserving' mechanism has so far been demonstrated in two species, the guinea pig and coypu, and it allows a marked increase in the blood level without necessarily an increased rate of secretion.

In many species progesterone is rapidly metabolized and removed from the blood. Investigations suggest that the steroid is distributed between two theoretical 'compartments' in the body; the rate of removal from the first compartment (possibly vascular) has a half-life of about a minute, whereas that of the second (possibly extra-vascular) about 30 minutes. In the pregnant guinea pig the presence of PBP with a high affinity for progesterone greatly decreases its rate of destruction. Progesterone production during gestation reaches a value of 200 μg/min (about 300 mg/day) in women and 1 μg/min (about 1.5 mg/day in guinea pigs. In both species the variation between individuals in production rates is considerable. These values are appreciably greater than the daily requirements for pregnancy maintenance.

Oestrogens. In most species there is an increasing level of oestrogens in blood and urine throughout pregnancy. In women this is partly attributable to the greatly elevated production of oestriol, derived mainly from the feto-placental unit. The total quantity of oestriol excreted in urine can be up to 50 mg/day, which is about ten times more than that of oestrone and oestradiol-17β. Few results are yet available from isotope studies for the production rates of oestrogens in pregnancy, but data for oestradiol-17β and oestrone indicate values of up to 30 mg/day. A large proportion of the oestrogens in circulation are conjugates; in contrast, progesterone is metabolized to more polar compounds before it is excreted in a conjugated form.

99

Role of hormones in pregnancy

A special case is that of the mare. In the second half of gestation when progesterone can no longer be detected in circulation, and the endometrial cups have ceased to secrete PMSG, there is a substantial increase in the oestrogens of maternal urine, in particular equilin, equilenin and oestrone. These oestrogens are probably produced by the feto-placental system, and they appear in high concentrations in maternal urine at a time when the fetal gonads undergo a conspicuous hypertrophy due to the rapid development of interstitial tissue. This species is a good example of one in which pregnancy is maintained by the local action of placental progesterone on the uterus, since circulating levels are extremely low though placental concentrations are notably high.

As in the case of progesterone, oestrogens are rapidly metabolized in pregnancy both by the liver and by extra-hepatic tissues in several species. This can be deduced from measurements of MCR, which may often exceed hepatic blood flow. However, oestradiol-17β is concentrated in target organs such as the uterus and mammary gland by specific tissue receptor proteins. Some recent evidence suggests that small quantities of progesterone may also be retained by a similar mechanism, even though a large proportion is rapidly metabolized by the target organ.

Cortisol and aldosterone. The secretory activity of the adrenals appears to be modified by the changing conditions of pregnancy. In the last trimester of gestation in women, the concentration of cortisol in blood reaches its maximum values, as does the secretion rate of aldosterone. The reasons for these increases, however, are quite different.

The increased blood levels of cortisol in pregnant women are related to a reduction in the rate of cortisol metabolism, whereas the actual rate of cortisol secretion decreases. The change in metabolism is largely attributable to a raised plasma concentration of corticosteroid-binding globulin (CBG), a protein produced by the liver, and one that possesses a high affinity for cortisol and corticosterone, as well as for progesterone. In the non-pregnant subject, the liver is the main site of corticosteroid metabolism, whereas in pregnancy, the increased levels of CBG result in a decreased liver metabolism and a reduction in the MCR of the steroid. The level of CBG appears to be related to the high concentrations of oestrogen found in gestation since non-pregnant patients treated with oestrogens show a comparable increase.

Other species differ in the kinetics of corticosteroid metabolism in pregnancy. In the rat, the MCR of corticosterone does not change but its secretion rate increases near term due to hyperactivity of the fetal and maternal adrenals. In sheep, the MCR of cortisol remains constant until the last 2 weeks of pregnancy, when it rises, probably because of a fall in CBG binding capacity.

Changes in maternal adrenal function are probably not directly related to the maintenance of pregnancy. We have already seen that some mammalian species may survive the stress of total adrenalectomy more successfully when pregnant or pseudopregnant than when non-pregnant. Furthermore, women with Addison's disease or patients bilaterally adrenalectomized may have normal pregnancies. The observed increase in aldosterone secretion in normal gestation may be due to an inhibition by progesterone of the action of mineralocorticoid on the absorptive mechanism of the renal tubule. An increased aldosterone secretion rate could be a compensatory response to the salt-losing action of high levels of progesterone in pregnancy.

Protein hormones. Little is known at present about the amounts of protein hormones required in pregnancy. Radioimmunoassay is providing information about their metabolic clearance rates and their production rates. Placental protein hormones such as HCG, HCS and PMSG are apparently produced in high concentrations and have long half-lives. Assays of HCG in urine give a production rate of 400 000 i.u./24 h, whereas PMSG, a gonadotrophin not found in urine, is produced at a rate of 200 000 i.u./24 h. The half-lives of HCG and PMSG are measured in hours (up to about 40 hr) and days (up to 6 days), respectively.

In contrast, pituitary protein hormones have a much greater clearance rate and a shorter half-life than placental protein hormones. Values for FSH, LH, GH and prolactin indicate that the half-life ranges from 5–30 minutes.

Role of steroid binding plasma proteins. An important property of certain plasma proteins is the transport of steroid hormones from the endocrine tissues where they are made, to the target tissues where they act. Some features of these steroid–protein interactions are given in Table 4-3. The role of these transport proteins is distinct from that of the binding proteins localized in target tissues which act as steroid receptors and are concerned with some of the primary events of steroid action.

Steroid hormones in blood are found in different physico-chemical forms; they may be bound to plasma proteins, conjugated with sulphuric and glucuronic acid residues, or present in a free form. An equilibrium exists between free and bound steroid; transport proteins have an important role as regulators of this dynamic state, and can influence the biological activity of hormones concerned in the maintenance of gestation.

The binding of steroid hormones to high-affinity plasma proteins provides a reservoir of steroid that dissociates only slowly. In the event of fluctuations in secretion, the time taken for the plasma concentration of a steroid bound to CBG to fall to zero is much longer than that of albumin-bound or unbound steroid. Another effect of high-affinity binding proteins is to protect a steroid against metabolism; the hepatic extraction of a bound steroid is greatly reduced and the MCR decreases conspicuously. In the guinea pig, a decrease in the rate of destruction of progesterone in pregnancy is associated with the appearance of high concentrations of PBP and CBG. A further role of transport proteins may be related to their ability to reduce the biological activity of a steroid by preventing it accumulating

TABLE 4-3. Plasma proteins binding and transporting steroids during pregnancy.

Plasma protein	Molecular weight	Steroids bound	Comments
Albumin (Human)	69 000	Androgens, corticosteroids, oestrogens, progestagens.	Low affinity and high capacity binding.
α_1-Acid glycoprotein, AAG (Human)	41 000	Progesterone, testosterone.	Medium affinity and medium capacity binding.
Transcortin, CBG (Human, monkey, guinea pig, rat and many other species)	52 000	Cortisol, corticosterone, progesterone.	High affinity, low capacity binding.
Sex steroid-binding protein, SBP (Human)	52 000–100 000	Oestradiol, testosterone.	High affinity, low capacity; increases in pregnancy.
Progesterone-binding protein, PBP (Guinea pig, coypu)	100 000	Progesterone	High affinity binding; concentration increases in pregnancy.

in the unbound form in extra-vascular spaces where its physiological effect may be deleterious. A possible example is to be found in women and guinea pigs, where increased amounts of transport proteins that bind testosterone may protect the mother and the fetus from the increased amounts of androgens produced in pregnancy.

So far as the maintenance of gestation in women is concerned, there are complex interactions between hormones and their plasma transport proteins, brought about by the changing endocrine conditions of pregnancy. The production of oestrogens, progesterone, PBP and CBG all increase. The increase of CBG is related to the high levels of oestrogens that are thought to stimulate the hepatic synthesis of this protein. However, not all species exhibit this degree of change in binding protein concentrations. It may be significant that the species in which an appreciable rise in CBG and PBP has been found in pregnancy have a haemochorial placenta, where maternal blood bathes fetal tissues, and they all show plasma progesterone levels which are much greater than those in the non-pregnant condition, as in women and guinea pigs (Fig. 4-5).

Pregnancy is essentially a partnership, at least from the hormonal point of view. The new individual is much more than just a parasite or passenger, for there is close integration between the fetal and maternal endocrine machinery which is essential for the continuance of pregnancy and may even control its termination. It is true that the mother's body supplies all the needs of the fetus, but from conception the new life plays an active part in safeguarding its future. The conceptus may signal to the mother its arrival in the uterus; produce the hormones that ensure its safe lodging; or furnish a stimulus which not only triggers birth, but in the end guarantees a source of food in the early days of life in the outside world by switching on a milk supply from the mother. It is in the regulation of these widely differing phenomena that hormones, acting as chemical messengers, play such a prominent role in the economy of gestation.

Suggested further reading

SUGGESTED FURTHER READING

The endocrine functions of the placenta. E. C. Amoroso and D. G. Porter. In *Scientific Foundations of Obstetrics and Gynaecology*, London; Heinemann (1970).

Sharp increase in free circulating oestrogens immediately before parturition in sheep. J. R. G. Challis. *Nature, London* 299, 208 (1971).

The four direct regulatory factors of myometrial function. A. I. Csapo. In *Progesterone: Its Regulatory Effect on the Myometrium. Ciba Foundation Study Groups*, No. 34 (1969).

The endocrinology of pregnancy and foetal life. R. Deanesly. In *Marshall's Phsyiology of Reproduction*, 3rd ed., vol 3. London; Longmans (1966).

Formation and maintenance of corpora lutea in laboratory animals. G. S. Greenwald and I. Rothchild. *Journal of Animal Science*. Suppl. 1, 27, 139 (1968).

Foetus and Placenta. Ed. A. Klopper and E. Diczfalusy. Oxford; Blackwell (1969).

Effect of embryo on corpus luteum function. R. M. Moor. *Journal of Animal Science*. Suppl. 1, 27, 97 (1968).

The Ovarian Cycle of Mammals. J. S. Perry. Edinburgh; Oliver and Boyd (1971).

Implantation and the maternal recognition of pregnancy. R. V. Short. In *Foetal Autonomy*. Ed. G. E. W. Wolstenholme and M. O'Connor. Ciba Foundation Symposium. London; Churchill (1969).

Hormonal maintenance in pregnancy. R. B. Heap, J. S. Perry and J. R. G. Challis. In *Endocrinology, Handbook of Physiology*. American Physiological Society (1972).

5 Lactation and its hormonal control
Alfred T. Cowie

The mammary gland is the distinguishing feature of mammals, for by definition these are animals that feed their young with milk secreted by such glands. The mammary gland is part of the reproductive apparatus, and lactation is the final phase of reproduction. In most mammals it is an essential phase, and failure to lactate, like failure to ovulate, means failure to reproduce. The stage of development at which the offspring becomes dependent on milk from its mother for nourishment varies greatly. In the lowly mammals – the monotremes – which have retained the reptilian practice of laying eggs, the young hatches from the egg when less than 2 cm long; it is a tiny fetus-like creature with fore-limbs that are well formed but with hind-limbs merely at the bud stage (Fig. 5-1a); of its sense organs probably only the olfactory system is functional. After hatching, the 'fetal' monotreme is entirely dependent for its further existence on its ability to suck and lick its mother's milk which is ejected from the underlying mammary glands on to a specialized area of skin – the areola – for the monotreme is without nipples. Somewhat higher in the evolutionary scale, the young marsupial, although born and not hatched, also enters the world in a 'fetal' state, for in most marsupials the embryo is nourished *in utero* exclusively through the yolk-sac, which has a much shorter functional life than the allantoic placenta of higher mammals (see Book, 2 Chapter 3). At birth only the fore-limbs are well developed (Fig. 5-1b) and with these this newborn mite mountaineers unaided from the birth canal to the pouch and hence to a teat which it grasps between its jaws and remains so attached until it is sufficiently grown to make excursions from the pouch; it will, however, continue to suckle regularly for some considerable time.

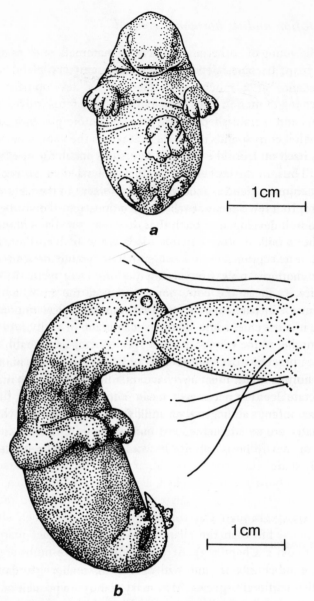

Fig. 5-1. *a*. The newly born echidna (Spiny Anteater).
b. The newly born kangaroo attached to the teat. Note in both species the well-developed fore-limbs, and that while the eyes are still covered with skin the nostrils are open. (From M. Griffiths, *Echidnas*. Oxford; Pergamon Press (1968).)

Lactation and its hormonal control

The young of eutherian or placental mammals, such as mouse and man, becomes dependent on the mammary gland for its sustenance at a much later stage of its development than the young of monotremes or marsupials; as a fetus it lives in the uterus and is nourished by way of an allantoic placenta so that at birth it corresponds to the stage at which the young marsupial casts itself off from the teat and leaves the pouch for brief intervals. Thus, in the eutherian mammal, the placenta has replaced the mammary gland as a source of nourishment in the early stages of life. In a few species, such as the guinea pig, the young may be so well developed at birth that they may survive without the mother's milk if other suitable foods are available. This, however, is exceptional and most eutherian young depend upon their mother's milk for varying periods after birth. In some species, such as the cow, horse and pig, the milk, and the colostrum in particular, is of great immunological importance since it is the chief route for the transmission of antibodies from the mother to the young; in other species, such as the rabbit and guinea pig, the main route seems to be by way of the placenta. The human infant must also have milk but failure of its mother to lactate does not necessarily mean failure to reproduce, for the human infant can survive on milk from other species. Not all neonates are so adaptable, and had cow's milk been first used to rear newly born rabbits it would have been regarded as highly toxic!

THE MAMMARY GLAND

Gross anatomy

The number, shape and size of the mammary glands vary greatly in different species. Mammary glands are present in both sexes (except in male marsupials) although they are frequently poorly developed in the male, and normally are functional only in the female. In most species the mammary glands are paired, varying from two, as in man, guinea pig and goat, to fourteen to eighteen as in the sow; in some marsupials the number is odd because two glands fuse at an early stage of fetal development. Some species have pairs of glands closely apposed in a structure termed an udder, like the two pairs in the cow, and one pair in the goat and sheep (Fig. 5-2). The position of the mammary glands varies – thoracic in man, elephant, monkey and the bat; extending along the whole length of the ventral thorax and abdomen in the sow, rat and rabbit; inguinal in the ruminants; abdominal in the whale and even dorsal in the coypu. As to shape and form there is great variety – in the rat the six glands on either side form relatively flat sheets of tissue enveloping the body wall, in the rabbit the glands are flat but circular in out-line; they may be prominent as in man or dependent as in ruminants. In all female mammals, with the exception of the monotremes, a nipple or teat is present on each mammary gland; in the males of some species, such as the rat and mouse, nipples are absent.

Whatever the external shape of the mature mammary gland may be, its basic internal structure is the same in all species. There are two distinct types of tissue within the mammary gland: first the true glandular tissue or parenchyma, and secondly the supporting tissue or stroma. The parenchyma in the func-tional gland consists of minute sac-like structures termed alveoli whose walls consist of a single layer of epithelial cells which are the milk-secreting cells (Fig. 5-3). The alveoli occur in clusters; each alveolus opens into a small duct and these small ducts join up to form larger ducts which eventually open to the exterior at the tip of the nipple or teat; in monotremes, as noted above, they open on special areas on the surface of the skin of the abdomen. While the basic structure of the mammary gland is similar, there is much species variation in the precise pattern of the duct system. In the rat and mouse the ducts eventually join to form one common duct or galactophore which leads directly through the nipple (Fig. 5-4a); in the rabbit the mammary ducts unite until there are some six to eight main ducts or galacto-phores, each draining a sector of the mammary gland, and these

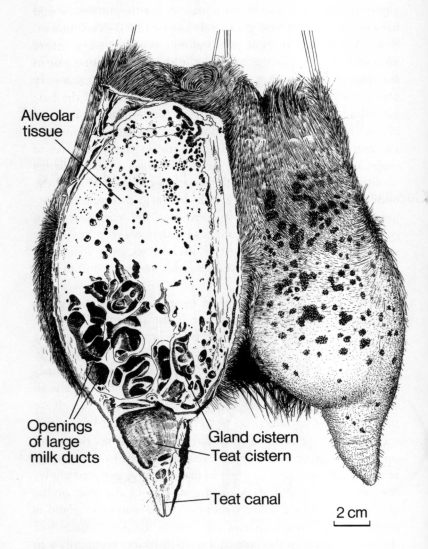

Alveolar
tissue

Openings
of large
milk ducts

Gland cistern
Teat cistern

Teat canal

2 cm

Fig. 5-2. Udder of goat. A portion of the left mammary gland has been removed to show the dense alveolar tissue, the gland cistern with the large ducts opening into it, the teat cistern, and the teat canal.

pass separately through the nipple (Fig. 5-4b). In some species certain parts of the ducts are dilated to form sinuses which act as storage spaces. In man, for example, some twelve to twenty galactophores pass through the nipple and each, in the region of the base of the nipple, expands to form a sinus (Fig. 5-4c). In

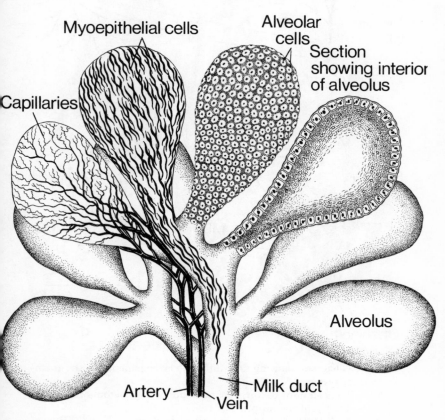

Fig. 5-3. Diagram of a cluster of alveoli in the mammary gland of a goat.

ruminants the mammary storage space is quite gross; the larger ducts terminate in a common gland cistern, a large cavity within the gland which leads directly into a smaller cistern within the teat; this in its turn leads to the teat or streak canal which opens at the tip of the teat (Fig. 5-4d).

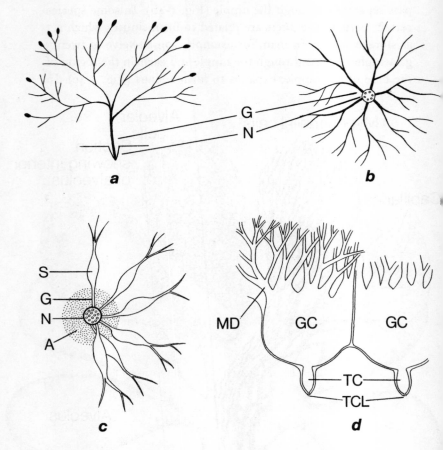

Fig. 5-4. Diagram showing four different arrangements of the mammary duct system.

a. Rat – ducts all unite to form one main duct (G, galactophore), which opens at the tip of the nipple (N).

b. Rabbit – ducts unite to form several main ducts.

c. Woman – each main duct is dilated near the base of the teat to form a sinus (S); the nipple is surrounded by the dark-coloured areola (A).

d. Ruminant – the main ducts (MD) open into a single large gland cistern (GC) which in turn opens into the teat cistern (TC). (TCL, teat canal.)

The varied positions of the mammary glands in different species mean that their blood supply and nervous connections must also differ. Since the mammary gland is of cutaneous origin it shares the blood and nerve supply of the contiguous skin.

Microscopic structure

In the fully developed functional mammary gland the alveolar walls are formed by a single layer of epithelial cells whose shape varies with the amount of secretion being stored in the lumen of the alveolus; when the lumen is empty the cells are tall, when the alveolus is full of secretion the cells are low and stretched (Fig. 5-5 *a* and *b*). Overlying the base of the epithelial cells is a network of star-shaped myoepithelial cells which, because of the manner in which they envelop the alveolus, were at one time termed basket cells (Fig. 5-6). On top of the myo-epithelium is a network of capillaries which supply the alveolar cells with the necessary precursor substances for the synthesis of milk.

When examined under the electron microscope (Fig. 5-7), the alveolar cells in the functional gland have on their free (i.e. luminal) surface numerous fine projections, the microvilli. Each cell is firmly joined to its neighbour by junctional complexes just below the luminal surface; the lateral surfaces of the alveolar cells are almost straight. The bases of the alveolar cells abut on the myoepithelial cells or on the basement membrane and are indented into a system of clefts which, by increasing the surface area of the cells, probably facilitate absorption of milk precursors. The cell nucleus is large and rounded. A characteristic feature of the cytoplasm is the abundant endoplasmic reticulum consisting of cytoplasmic membranes arranged as arrays of flattened sacs orientated parallel to one another with their outer surfaces covered with numerous RNA protein particles (ribosomes) and mostly situated in the basal two-thirds of the cell. The mitochondria are large with conspicuous partitions, or

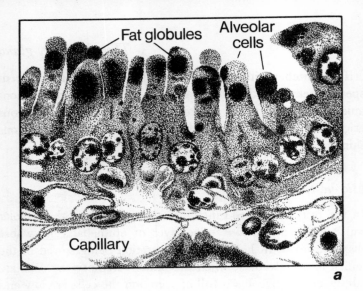

Fat globules | Alveolar cells

Alveolar cells

Capillary

a

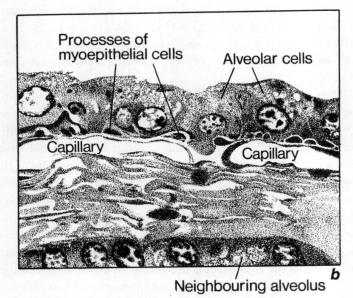

Processes of myoepithelial cells

Alveolar cells

Capillary

Capillary

b

Neighbouring alveolus

Fig. 5-5. Cells in the wall of an alveolus, (*a*) just after milking, (*b*) just before milking. As the alveolus fills up with milk and its walls are stretched the shape of the cells is much altered. The empty capillaries at the base of the alveolar cells show up clearly because the tissue was fixed by intravascular perfusion. ((*a*) from S. J. Folley. In *Marshall's Physiology of Reproduction*, 3rd ed. vol. 2, Chap. 20. Ed. A. S. Parkes. London; Churchill (1952). (*b*) from K. C. Richardson, *Proc. R. Soc. B*, **136**, 30 (1949).)

cristae. Another striking ultrastructural feature of the functional alveolar cell is the large Golgi apparatus, located in the more apical part of the cell, also adjacent to the nucleus, consisting of

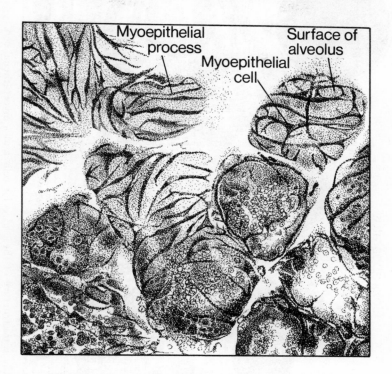

Fig. 5-6. Myoepithelial cells on the outer surface of alveoli. (Photograph by courtesy of Mr K. C. Richardson.)

stacks of flattened sacs and vacuoles which, by their numbers, may give sections of that part of the cell a 'moth-eaten' appearance. The myoepithelial cells have a spindle-shaped nucleus, their cytoplasm is rarefied and contains fine muscle-like myofilaments which run parallel to the long axis of the cell's processes and are attached to the cell membrane. As noted later these myoepithelial cells have a contractile function and are concerned in the expulsion of milk from the lumen of the alveolus into the duct system.

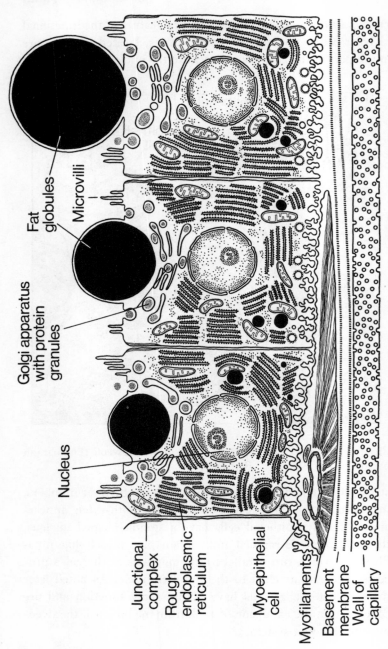

Fat globules

Microvilli

Golgi apparatus with protein granules

Nucleus

Junctional complex

Rough endoplasmic reticulum

Myoepithelial cell

Myofilaments

Basement membrane

Wall of capillary

Fig. 5-7. Diagram of the ultrastructure of three alveolar cells and a myoepithelial cell.

Growth of the mammary gland in the fetus

The evolution of the mammary gland has been much discussed and its origins variously associated with the primordia of sweat glands, sebaceous glands, hair follicles or avian incubation patches, but I do not propose to extend such speculations. The mammary glands are derived from the ectoderm in the embryo: two linear thickenings appear as a ridge, the milk line or crest, on either side of the mid line. This ridge becomes interrupted

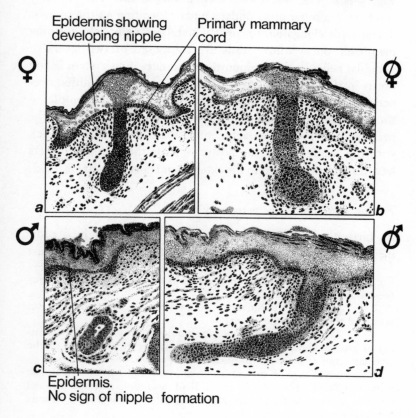

Fig. 5-8. Development of the mammary gland in the fetal mouse and the effect of gonadectomy. (From A. Raynaud. In *Milk: the Mammary gland and its Secretion*, vol. 1, Chap. 1. Ed. S. K. Kon and A. T. Cowie. New York and London; Academic Press (1961).)

into a series of nodules of ectodermal cells, the number and position of these depending on the species. These nodules, the early rudiments of the mammary glands, sink into the dermis to become mammary buds. At first lens-shaped, the bud becomes more spherical and later conical. A pause then ensues in the bud's growth after which it elongates into a cord-like structure, the primary mammary cord, whose base remains attached to the epidermis while its distal end penetrates into the dermis (see Fig. 5-8a). The distal end then branches into two, or may produce a number of secondary buds depending on the species. These buds elongate into cords which become hollow, forming mammary ducts whose walls are composed of two layers of cuboidal cells. In this fashion the mammary duct system is laid down in the developing fetus. In some species sex differences occur in the pattern of fetal mammary growth; for example, in the male rat and mouse in the second half of fetal life the primary cord loses its attachment to the epidermis (Fig. 5-8c), with the result that in the postnatal male the mammary ducts have no communication to the exterior and no nipples are formed.

Postnatal growth of the mammary gland

At birth the mammary apparatus is represented by a rudi- mentary duct system leading to a small nipple, and until just

---

Fig. 5-9. Sections of the mammary gland of the goat during pregnancy (gestation is about 150 days):

    35th day of pregnancy – note the small collections of ducts scattered throughout the stroma;

    92nd day of pregnancy – the lobules of alveoli are now form- ing in groups known as lobes; secretion is present in some of the alveolar lumina; there is still quite a lot of stromal tissue;

    120th day of pregnancy – the lobules of alveoli are almost fully developed; the alveoli are full of secretion and the stromal tissue is reduced to thin bands separating lobules and thicker strands between lobes.

(*35th day* and *120th day* from A. T. Cowie. In *Lactation, Proceedings of an International Symposium forming the Seventeenth Easter School in Agricultural Science, University of Nottingham, 1970*. Ed. I. R. Falconer. London; Butterworths.)

Duct   Stroma   Stroma Lobule Alveolus

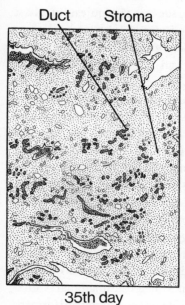

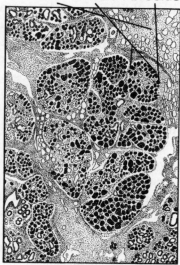

35th day   92nd day

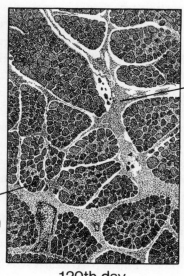

Stroma
separating
lobes and
lobules
of alveoli

Alveolus
filled
with
secretion

120th day

before the onset of reproductive cycles in the female, the mammary duct system grows but little, for this is a quiescent phase in its development. As the onset of regular ovarian cycles approaches, a phase of active mammary growth begins, the nature of which varies with the species and is related to the type of sex cycle. In species with short cycles, e.g. of 4–5 days as in the rat and mouse, the ducts grow and branch rapidly until they are fully extended, but alveoli are not formed. In species with a more prolonged luteal phase to the cycle, such as man and monkey, branching of the duct system continues until the future lobules are indicated by collections of fine ductules and even alveoli surrounded by delicate stromal tissue; thus in these species a more pronounced growth of the mammary gland occurs in the non-pregnant animal than in those with short cycles. Lastly, in those species in which the luteal phase of the cycle is prolonged into a pseudopregnancy the duct growth and lobulo-alveolar formation proceed to a degree that is observed only in late pregnancy in the first two groups.

In some species there is also considerable growth of the stromal and supporting tissues of the mammary gland during this period, as evidenced by the growth of the breast in man and of the udder in ruminants. But in the non-pregnant female the overall size of the breast or udder may not accurately reflect the amount of true glandular tissue within.

In most mammals full mammary growth is not achieved until the end of pregnancy or even early lactation. Usually not until mid pregnancy (Fig. 5-9) is the full impetus of mammary growth observed; then there is a rapid formation of lobules of alveoli which take over much of the space formerly occupied by stromal tissue, so that by the last third of pregnancy the stroma is represented by narrow bands of connective tissue further dividing the gland into lobes. In some species during the last third of gestation the alveolar cells begin to secrete and the lumen of the alveolus becomes distended with a secretion containing fat globules. This is particularly striking in ruminants (Fig. 5-9); in other species, such as rats, secretory activity is not observed until just before parturition. In all species there is a further burst of secretory activity at the time of parturition or soon afterwards.

Control of fetal mammary growth

Only in the rat and mouse have we as yet any information about the factors regulating mammary growth in the fetus. In these species from the beginning of the last third of gestation, the growth of the fetal gland is influenced by gonadal hormones, the separation of the primary mammary cord from the epidermis and the suppression of the nipple in the male fetus being brought about by androgens from the fetal testis. If the testes are destroyed before this period the gland will develop as in the female (Fig. 5-8d), i.e. it will remain attached to the epidermis and a nipple will be formed. Destruction of the ovaries of the female fetus, on the other hand, does not alter mammary growth (Fig. 5-8b), but injections of androgens into the female fetus induce the male pattern of mammary growth. Thus in the fetal rat and mouse the male pattern of mammary growth is of a specialized type, produced by hormones from the fetal testis which modify the neutral (female) pattern of growth. Whether such hormone-mediated alterations of fetal mammary growth occur in other species is not known; the question may be of some importance because some of the so-called 'spontaneous' malformations of the mammary gland of the newborn may be associated with the effects of hormones administered to the mother during pregnancy.

In the earlier stages of mammary growth in the fetus, the underlying mesenchyme influences the differentiation of the mammary rudiments by exerting an inductive function on the epidermis through organ-specific factors; i.e., it is the mesenchyme that determines whether the overlying ecto-dermal cells become organized into a salivary or a mammary gland.

121

Control of postnatal mammary growth

Little if anything is known about the control of postnatal mammary growth in either monotremes or marsupials, but it has been widely studied in some placental mammals. In these, mammary growth is largely controlled by hormones arising from the anterior pituitary gland, from the ovaries, and from the adrenal cortex; during pregnancy, the placenta may be an additional

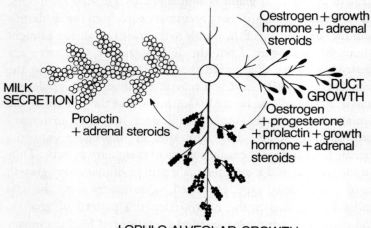

Fig. 5-10. The hormones concerned in the growth of the mammary gland and in the initiation of milk secretion in the hypophysectomized-ovariectomized-adrenalectomized rat. (From W. R. Lyons, *Proc. R. Soc.* B, **149,** 303 (1958).)

source of both steroid and polypeptide hormones, but the relative importance of certain of these hormones appears to vary with the species. Of all the body organs none is more complexly regulated by hormones in its growth and function than is the mammary gland.

In the rat and mouse, oestrogen from the ovary triggers off the sudden spurt of mammary duct growth just before the onset of oestrous cycles; ovariectomy abolishes the onset of this increased growth, which can then be restored by injection of suitable doses of oestrogen. Growth of the lobulo-alveolar system usually requires both oestrogen and progesterone – produced by the ovaries, by the adrenal cortex or by the placenta – although in some species oestrogen alone, if given in suitable doses, will induce lobulo-alveolar growth. This, however, is only part of the story, for these steroid hormones have little or no effect on mammary growth if the anterior pituitary is removed. Researches show that the steroids affect mammary growth in two ways, first by releasing from the anterior pituitary prolactin and probably other hormones, and secondly by acting directly on the mammary parenchyma, but the latter action is possible only if anterior pituitary hormones are also present. By removing the ovaries, adrenals and pituitary from infantile rats, the minimal hormonal requirements for mammary growth in these 'endocrinectomized' animals can be determined by studying the effects of various combinations of hormones. For duct growth, oestrogen, growth hormone and adrenal steroids are necessary; lobulo-alveolar growth will ensue if to the above triad prolactin and progesterone are added (Fig. 5-10). To what extent these observations apply to other species is still uncertain, but in the hypophysectomized goat a recent preliminary study has suggested that all five hormones are required for lobulo-alveolar growth. Prolactin-like hormones are present in the placenta of primates, goats and rats, and may well occur in other species, and these placental hormones could have important roles in mammary growth during pregnancy.

LACTATION

From substances taken up from the blood the alveolar cells synthesize within their cytoplasm the constituents of milk and then pass them out into the lumen of the alveolus. While being stored within the alveolus some further changes in the composition of the secretion may take place, since fluctuations in the osmotic pressure of the blood traversing the neighbouring capillaries may cause a transfer of water and water-soluble constituents between alveolus and capillaries. These processes together make up what is termed milk secretion – the first phase of lactation. The second phase is known as milk removal, and is the process whereby the secreted milk stored in the alveoli and fine ducts is moved towards the large ducts, sinuses, or cisterns according to the species, and thence to the teat or nipple where it becomes available to the suckling infant or the milker.

Cytology of milk secretion

The cellular mechanism of milk secretion has been a source of argument for over a century. Some microscopists maintained that milk was formed by the disintegration or fatty degeneration of the alveolar cells, or by a process involving cell decapitation and the extrusion of the cell's contents directly into the lumen. Others maintained that the milk constituents were passed out through the cell membrane with little loss of cytoplasm, but the size of the fat droplets made such a process difficult to visualize. Not until the advent of the electron microscope with its high resolving power was a credible solution forthcoming. It now appears that the fat is synthesized within the endoplasmic reticulum in the basal portion of the cell (Fig. 5-7) and the small membrane-bound fat droplets move towards the apex of the cell and increase in size. On reaching the apex the fat droplet gradually pushes against the cell membrane which loses its microvilli and bulges into the lumen. As this process continues, the cell membrane fits tightly over the droplet and gradually constricts behind it so that a narrowing neck of cell membrane is formed between the enveloped droplet and the cell apex; this neck narrows till its membranous walls fuse, and the fat droplet, now completely enclosed in a double cell membrane, is pinched off and drops free into the alveolar lumen while the cell apex remains intact. Not infrequently, however, a portion of cytoplasm may become pinched off and be enveloped with the fat droplet.

The milk protein first appears as fine granular material within the vacuoles of the Golgi apparatus (Fig. 5-7); these vacuoles and their granular contents move towards the cell apex; they come in contact and eventually fuse with the cell membrane which then ruptures thus allowing the contents to be discharged into the lumen. Recent studies suggest that lactose is also formed within the Golgi apparatus in association with the milk protein; indeed there is an interesting relationship between lactose and protein. The final step in the biosynthetic pathway of lactose is accomplished by an enzyme called lactose synthetase, which is made up of two protein components one of which is the milk protein α-lactalbumin. α-Lactalbumin is believed to be synthesized within the membranes of the rough endoplasmic reticulum and then passed into the Golgi apparatus where the second component of the enzyme – galactosyltransferase – is probably attached to the Golgi membranes. The conjunction of the two components of lactose synthetase thus permits the completion of the synthesis of lactose which then passes into the lumen, probably with the protein granules. The synthesis of lactose is thus dependent on the prior synthesis of the milk protein α-lactalbumin.

Hormonal control of milk secretion

As already noted milk secretion begins in some species during the last third of gestation, in others the onset of secretory activity is more closely associated with parturition. The secretion formed at this time is called 'colostrum' and differs in composition from normal milk. Soon after parturition there is a greatly increased secretory activity, and the colostral type of secretion changes to the formation of ever increasing amounts of milk of normal composition.

How is the onset and the maintenance of secretion controlled? Various studies, including some striking experiments by Jim Linzell in goats in which the mammary gland had been transplanted to the neck without subsequent loss of its ability to secrete milk, clearly indicate that secretory nerves are not

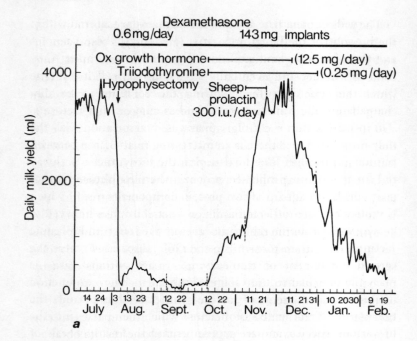

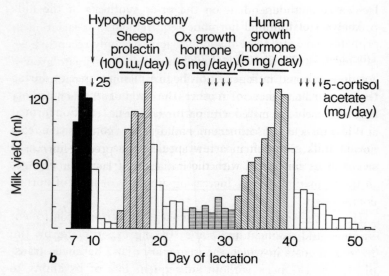

concerned and that the control mechanisms are hormonal. As in the control of the growth of the mammary gland, so the anterior pituitary gland plays an all-important role in its function. In the few species so far studied, surgical removal (hypophysectomy) or destruction of the anterior pituitary causes an immediate decline and ultimately a cessation of milk secretion. The precise anterior pituitary hormones concerned, however, may vary. In the lactating rat, after removal of the anterior pituitary, prolactin and adrenocorticotrophin are both essential for maintaining milk secretion; in the hypophysectomized goat, prolactin, growth hormone, adrenocorticotrophin and thyrotrophin are all required for full milk secretion (Fig. 5-11a), whereas in the rabbit, after removal of the pituitary, prolactin alone can restore milk secretion (Fig. 5-11b). Studies carried out *in vitro* on mammary glands or portions of mammary glands explanted into artificial culture media to which hormones are added, are now providing much additional information on the direct effects of hormones on mammary function in various species, and are supplementing the results obtained by *in vivo* methods.

Analyses of minimal hormone requirements for lactation in primates have still to be made. Prolactin has only recently been isolated from human and monkey pituitaries. Primate growth hormone exerts lactogenic (milk-stimulating) properties in rabbits similar to those of ruminant prolactin and this hormone may have a role in milk secretion in primates.

While there is thus information on the hormones necessary to sustain milk secretion in a few species, the precise hormonal mechanisms concerned with the initiation of lactation are less

Fig. 5-11. Diagrams of daily milk yields of (a) goat and (b) rabbit, hypophysectomized during lactation and then treated with hormones to restore lactation. ((a) from A. T. Cowie. In *Lactogenesis: the Initiation of Milk Secretion at Parturition*. Ed. M. Reynolds and S. Folley. University of Pennsylvania Press, Philadelphia (1969). (b) from A. T. Cowie, P. E. Hartmann and A. Turvey, *J. Endocr.* **43**, 651 (1969).)

Lactation and its hormonal control

well established. The differences in the time of onset of secretory activity in the mammary gland in different species, moreover, would suggest that several mechanisms may exist. We have seen that certain anterior pituitary hormones, such as prolactin, growth hormone and adrenocorticotrophin, have roles both in

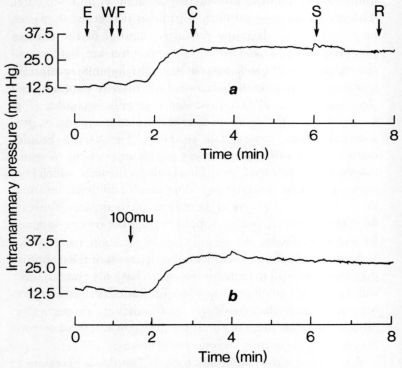

Fig. 5-12. Intra-mammary pressure recorded from an unmilked hind quarter of a cow (*a*) during machine milking and (*b*) during a rapid intravenous injection of 150 mμ oxytocin.
E, entry of milker; W, washing of udder; F, fore milking; C, application of teat cups; S, stripping; R, removal of teat cups. J. D. Cleverley and S. J. Folley, *J. Endocr.* **46,** 347 (1970).

mammary growth and milk secretion, and the question that arises is – how is their action switched from one of stimulating mammary growth during pregnancy to that of stimulating mammary function in late pregnancy or after parturition? Only

128

in the rat, and to a lesser extent in the mouse, do we have experimental evidence about this mechanism. In the pregnant rat the lactogenic activity of pituitary prolactin and placental lactogen is blocked by a direct action of progesterone on the alveolar epithelium. Some thirty hours before parturition there is a fall in the steroid production by the ovaries and a resulting drop in the level of progesterone in the blood which permits the lactogenic effects of prolactin, placental lactogen and adrenal steroids to come into play. In the rat and mouse there is some evidence that progesterone exerts part of its inhibitory effect by blocking the synthesis of α-lactalbumin, thereby inhibiting the synthesis of lactose. In other species other mechanisms have been suggested for the onset of milk secretion, such as sudden increases in blood levels of active adrenal steroids, increased levels of anterior pituitary hormones in response to reduced levels of steroids following loss of the placenta, or in response to the suckling stimulus itself.

Milk removal

We can now consider the second phase of lactation, namely milk removal, which is concerned with the transport of the milk from the alveolar lumina to the nipple or teat where it becomes available to, and can be removed by, the suckling infant or the milker. This process involves the triggering of a neurohormonal reflex, the milk-ejection reflex.

People have long known that a few minutes after the infant is put to the breast, or after the milker starts to milk the cow, the mammary gland seems suddenly to fill up with milk, which comes under pressure and might spurt from the nipple or teat (Fig. 5-12). This phenomenon was called the 'draught' in women and the 'let-down of milk' in cows. It could, moreover, occur in anticipation of the suckling or milking stimulus. On the other hand, discomfort or any form of stress can stop the occurrence of the 'draught' or 'let-down' so that much less milk is obtained by the child or by the milker. For many years the sudden milk flow was attributed to an increase in the rate of milk secretion. However, the udder of a cow was then shown to contain all the milk obtainable on milking *before* the occurrence of the 'let-down', and there is now ample evidence that the rapid increase in intramammary pressure is due to the sudden expulsion of the milk from the alveoli and fine ducts into the large ducts, sinuses or cisterns. This movement of milk is brought about by contraction of the myoepithelial cells squeezing the alveoli and expelling their contents. The phenomenon is now termed milk ejection. Extensive investigations have revealed that the act of suckling or milking triggers nerve impulses, from receptors in the nipple or teat, which pass up the spinal cord to the hypothalamus where they cause the release of the hormone

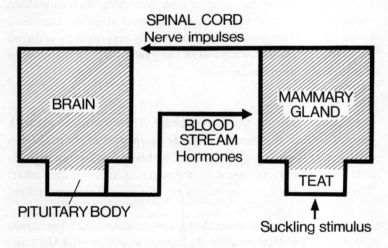

Fig. 5-13. Diagram to indicate relationship between mammary gland and the brain. The act of suckling triggers off nerve impulses which pass from the teat or nipple to the spinal cord and thence to the brain. Impulses pass through the hypothalamus (i) to the posterior pituitary causing the release of oxytocin into the blood which, when it reaches the mammary gland, induces contraction of the myoepithelial cells and milk ejection; (ii) to the regions controlling the anterior pituitary and so lead to the release of anterior pituitary hormones, including prolactin, necessary for milk secretion. (From J. S. Tindall and G. S. Knaggs. In *Hormones and the Environment*, Memoirs of the Society for Endocrinology, Number 18. Ed. G. K. Benson and J. G. Phillips (1970).)

oxytocin from the posterior lobe of the pituitary gland into the blood stream (Fig. 5-13). On reaching the mammary gland oxytocin causes the myoepithelial cells to contract thus increasing intramammary pressure (Fig. 5-12). The milk ejection response is a reflex, but unlike ordinary reflexes its efferent arc is not nervous but hormonal. Like ordinary reflexes, however, it can be conditioned, which explains the occurrence of milk ejection before milking in response to stimuli associated with nursing or milking. The importance of the proper functioning

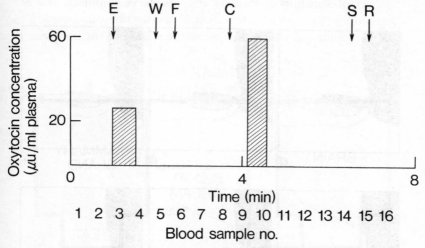

Fig. 5-14. Oxytocin levels in jugular vein blood of a cow during milking; note initial release of oxytocin in response to entry of the milker and a further release just after the application of the teat cup. (See Fig. 5-12 for explanation of the lettering.) (From J. D. Cleverley and S. J. Folley, *J. Endocr.* **46,** 347 (1970).)

of this reflex seems to vary with species; in the rat and rabbit the sucklings will get practically no milk even from mammary glands full of milk unless milk ejection occurs; at the other extreme one can milk out the udder of the goat even in the total absence of the milk-ejection reflex. This divergence may well reflect differences in the architecture of the mammary gland; there are no large storage sinuses or cisterns in the mammary glands of the rat or rabbit whereas there is a large gland cistern

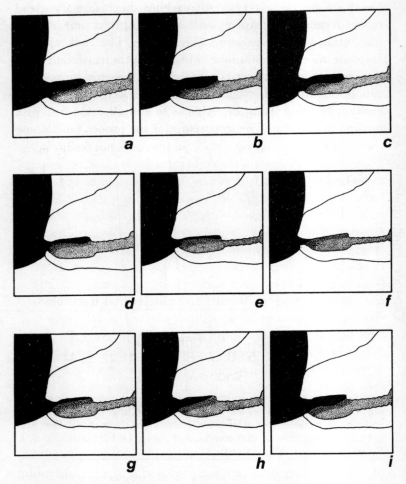

Fig. 5-15. Tracings from prints from a cineradiographic film of a kid feeding from the udder; a quantity of sterile barium sulphate suspension was injected into the gland cistern so as to render the milk opaque to X-rays:

a The teat cistern is full of milk – note the indentation produced on the lower surface of the teat by the tongue.

b and *c* The tongue in the region of the cheek teeth is lowered and there is a raising of the lower jaw and the tip of the tongue with a narrowing of the neck of the teat.

in the goat into which secretion may more readily pass. In some species, moreover, the myoepithelial cells can make at least a partial response to direct mechanical stimulation of the mammary gland which may aid milking.

In the modern dairy cow the release of oxytocin and milk ejection generally occur soon after the teat cups are applied; indeed they may occur before the application as a conditioned response to being fed before milking or to the sight or sound of the approaching milker (Fig. 5-14). Primitive breeds of cattle are less accommodating and milk ejection does not readily occur in them in the absence of the suckling calf. To aid in the milking of his domestic animals man has, in the past, adopted two main subterfuges which exploit certain aspects of the milk-ejection reflex. The obvious trick of bringing the calf to the cow and even allowing it to suckle one teat while the others were milked was frequently used; sometimes the presence of a dummy calf or even of a boy covered with a calf's skin might be adequate to elicit the reflex. The other and more strange procedure was to elicit milk ejection by stimulation of the genital tract, blowing air into the vagina being a common method of achieving this. The genital-stimulation procedure was apparently known to the ancient Egyptians, while Herodotus described how the Scythians used pipes made of bone to blow into the vagina of their mares before milking. Indeed the practice of blowing air into the vaginae of cows to aid milking has been widely used in Europe, Asia and Africa and is still used by primitive peoples in the latter continent. Although odd, the practice has a sound

---

d and *e* Further raising of the forepart of the tongue has occluded the neck of the teat and the contents of the teat cistern are being expressed into the back of the mouth.

f The final phase of expression.

g, h and *i*. Lower jaw and tongue are now being lowered to allow refilling of the teat.

a to *f*, and *g* to *i* are consecutive prints (50 frames per second). There is a 3-frame interval between *f* and *g*.

(From G. M. Ardran, A. T. Cowie and F. H. Kemp, *Vet. Rec.* **69**, 1100 (1957).)

physiological basis, for there is now clear evidence that mechanical stimulation of the female reproductive tract can cause the release of oxytocin from the posterior pituitary and thereby evoke milk ejection. The widespread use in several continents, and over several millenniums, of this technique to facilitate milking may at first sight seem remarkable and puzzling. Common, indeed essential, to all societies is the habit of copulation and no doubt primitive man, observing on occasions the spurting of milk from the nipples of a lactating partner during coitus, turned the thought over in his mind that if vaginal stimulation induced 'the draught' in women it might also induce 'let-down' in cows and mares. Modern man having similar preoccupations has now revealed the underlying physiological mechanism and has even detected oxytocin in the blood after female orgasm.

There exists a widespread belief that milk expulsion in marsupials and in marine mammals is aided by voluntary contraction of adjacent body muscles. In some specific instances this has now been shown to be false and there would seem to be good reasons for believing that the neuroendocrine reflex of milk ejection occurs in most if not all mammals including the monotremes, and that it is the principal mechanism for the transfer of milk from the alveoli to the large ducts and sinuses.

Physiology of suckling

Most people think that the young obtain milk by sucking the nipple or teat, but *they do not*. Suction aids the process, but it is not the essential component. Cineradiographic studies in young babies and animals during suckling clearly reveal that milk is obtained by expressing it from the nipple or teat (Fig. 5-15). The base of the teat is compressed between the tongue and hard palate, and the milk contained within is then stripped out of the teat by the tongue compressing the teat from its base towards its tip against the hard palate. The pressure on the base of the teat is then released to allow it to fill up with milk, which quickly happens as the milk is under pressure from the milk-ejection reflex, and the whole action is repeated. The human infant forms a 'teat' by drawing the whole nipple and part of the areola into its mouth but otherwise the procedure is the same. Young animals, however, very soon acquire the technique of obtaining milk by suction if fed from bottles with hard teats but this is not their normal method.

The suckling and milking stimulus

In the last decade rapid advances have been made in discovering the pathways within the spinal cord and brain used for the transmission of stimuli from the mammary gland to the hypothalamus; the spinothalamic system is mainly concerned. The final link in the oxytocin-release path is the paraventricular nucleus in the hypothalamus and its neurosecretory fibres which pass into the posterior lobe of the pituitary. The oxytocin so secreted into the blood can be readily measured by modern assay techniques (see Fig. 5-14).

In addition to bringing about the release of oxytocin required for milk ejection, the suckling or milking stimulus also releases prolactin from the anterior pituitary, and probably the other anterior pituitary hormones required for the maintenance of milk secretion. At present less is known about the nervous pathways involved, but a common path may be shared as far as the mid-brain. In contrast to its role in the control of other pituitary trophic hormones, the hypothalamus exerts a general inhibitory effect on the secretion of prolactin, so that if the pituitary be isolated from the hypothalamus by surgical transection of its stalk, or by transplantation, the secretion of prolactin continues and may even increase whereas the production of other hormones is much reduced. This inhibitory effect is brought about by the production of a neurohumour in the hypothalamus – not yet isolated but termed the prolactin-inhibiting factor – which is released into the hypophysial portal system which carries blood direct from the hypothalamus to the anterior pituitary. The recent development of sensitive radioimmunoassays for prolactin has made it possible to detect the rapid rises in blood prolactin which occur in response to milking in the cow and goat, and to suckling in the rat. The suckling or milking stimulus, being generally necessary both for milk secretion and for milk ejection, is thus of great importance in the general maintenance of lactation. In some species indeed, such as the cat and rat, it appears to be essential; however, in the goat, at least under experimental conditions, it is apparently dispensable, for in that species lactation can continue when all nervous connections between mammary gland and central nervous system are destroyed. Other factors must therefore affect the release of anterior pituitary hormones; for example changes in blood composition, brought about by the uptake of milk precursor substances by the mammary gland, may be detected by special centres in the hypothalamus thus effecting a metabolic release of prolactin. There is, moreover, evidence now that areas of the brain much 'higher' than the hypothalamus participate in controlling the release of anterior pituitary hormones so that the control of prolactin secretion may be of far greater complexity than hitherto believed.

CLINICAL STUDIES AND USES OF HORMONES IN LACTATION

Hormone preparations have been used both in man and in animals to initiate, to increase or to suppress lactation.

Induction of mammary growth and lactation

In the late 1930s when synthetic oestrogens, such as diethylstilboestrol, became available, the feasibility of using these to stimulate udder growth in ruminants was investigated, and these studies were extended during the Second World War with the object of bringing sterile cows and heifers into lactation. The usual procedure was to implant tablets of oestrogen beneath the skin from which the oestrogen was gradually absorbed. A few weeks after implantation 'milking' was started. Initially only a few drops of secretion were obtained but gradually the volume increased and within a month the daily milk yield might be about 5 kg. After 6–10 weeks the unabsorbed portions of the tablets would be removed when further increases in yield might occur so that from animals which responded successfully 13–14 kg of milk/day might be obtained. However, about half the number of animals treated failed to give a satisfactory response and because of this and other troublesome side effects, such as persistent oestrous behaviour, the procedure was never widely used. Subsequently, when synthetic progesterone became available, numerous studies were made on the induction of udder growth in goats with oestrogens and progesterone. The combination of hormones stimulated a type of lobulo-alveolar growth which was more normal in structure than that observed with oestrogen alone, when the alveoli had tended to be unduly large and even cystic. In virgin goats treated with oestrogen + progesterone to induce lactation, milk yields obtained were about two-thirds of those expected had the animal been pregnant and come normally into lactation. These studies in goats also revealed the important contribution of the milking stimulus in the mammary growth process. Indeed, virgin goats can be brought into lactation by regular 'milking' without giving either oestrogen or progesterone. These experimental observations are, of course, in line with the numerous clinical reports of lactation occurring in non-pregnant women in response to repeated application of the suckling stimulus. As already noted, the ovarian hormones exert much of their mammary-growth effects by releasing the appropriate hormones from the anterior pituitary, and the milking stimulus clearly enhances this release. Seldom, however, can full growth of the mammary gland comparable to that occurring in late pregnancy be obtained with oestrogen–progesterone therapy, probably because of the absence of placental hormones in such studies.

Ovarian hormone therapy, including the local application of the hormones in suitable cream, is sometimes used in the treatment of inadequate breast development in women, but unless growth of the stromal as well of the glandular tissue is stimulated the benefits obtained are usually slight. There may also be undesired side effects such as disturbance of the menstrual cycle; even more bizarre was the recent report of a husband who developed gynaecomastia because his wife had applied the hormone cream to herself either somewhat too lavishly or had failed to rub it in thoroughly. This report is not without its ironical twist as the treatment was started because the husband was dissatisfied with his wife's 'miserable little breasts' – it illustrates moreover the local action of the steroid hormones on the mammary gland and that in some species the male mammary gland can respond quite readily to mammogenic hormones!

Augmentation of an existing lactation

The milk yields of cows in declining lactation can be increased by suitable hormone therapy. Bovine prolactin would appear to be the obvious choice but it is ineffective for this purpose, the likely explanation being that there is no deficiency of prolactin during the decline of lactation in the cow. Bovine growth hormone, on the other hand, is effective, and the increases in milk yield are related to the dose of growth hormone administered over a fairly wide dose range (Fig. 5-16). However, until synthetic growth hormone becomes available this therapy cannot be used on a commercial scale.

Thyroid hormones have been used commercially to augment the milk yield of cattle. These, being active by mouth, can be incorporated into the food and give considerable increases in milk yield, but they must not be given for any length of time because of overstimulation of the general body metabolism; unfortunately when the treatment is stopped there is a dramatic acceleration in the rate of decline in milk yield which reduces the overall gain to a mere 5 per cent. Thyroid hormone therapy is now little used in cows.

Preparations of ruminant prolactin have in the past been

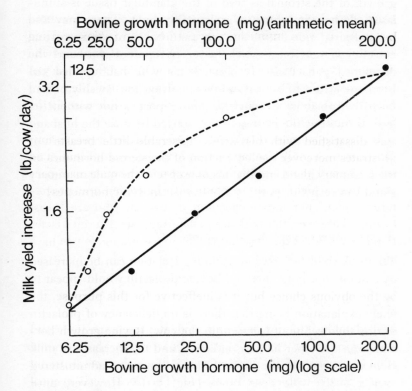

Fig. 5-16. Effect of increasing doses of bovine growth hormone on the milk yield of cows. Upper curve represents doses plotted on an arithmetic scale, and lower curve doses plotted on logarithmic scale. (From J. B. Hutton, *J. Endocr.* **16,** 115 (1957).)

used clinically to treat hypogalactia in women but without success; this is hardly surprising because prolactins may differ in different species and the prolactin from one species may be inactive in another. Indeed the use of heterologous prolactin may well be dangerous because of serious immune responses. Human growth hormone has recently been used with some suggestion of success. The problem of hypogalactia in women is complex and is discussed later.

Lactation and its hormonal control

Inhibition of lactation

Oestrogens have been used extensively to suppress unwanted lactation in women in the puerperium. Since the suckling stimulus also ceases in such cases, the mode of action of the hormone is still a matter for discussion and indeed the efficacy of the treatment has been questioned. More recently there have been doubts about its safety in view of a possible association with thrombo-embolism.

In certain species oestrogen + progesterone combinations can depress or inhibit lactation even in the presence of continued suckling, and the question has been raised whether oral contraceptives which usually contain both oestrogen and progesterone might not impair lactation in women who are breast feeding. However, the evidence to date suggests that once lactation is established the dose levels now in use are too low to have any deleterious effects.

Human lactation

Considerable attention is being paid at present to the rapid decline in breast feeding that is occurring in the so-called affluent societies of the Western world, in which at present only 20 per cent of babies are breast-fed. The decline has accompanied the striking improvements in artificial feeding; no longer are there marked differences in mortality between breast- and bottle-fed babies and the mother can therefore freely choose how she will feed her baby. By some this situation is regarded as a satisfactory step forward in the emancipation of women, by others a serious failure of normal lactation which may endanger the child both nutritionally and psychologically.

It can hardly be regarded as a physiological or endocrine failure of lactation when a mother freely decides to bottle-feed her offspring and to have her lactation suppressed. Some women, for example, may find that breast feeding interferes with their work or social routines while others may consider that lactation may be detrimental to their figures. However, many babies are perforce bottle fed because lactation has not been satisfactory. There has been considerable clinical research into the causes of such failures and numerous social and psychological factors have been implicated. These interrelated factors can be grouped under two main headings: first, individual emotions and attributes and, secondly, group-derived emotions and attitudes. In the first category, a mother who freely expresses a preference for breast feeding is more likely to succeed than one who expresses no preference. Maternal instinct also plays a role in that the decision to bottle feed is highly correlated with mother-centred reasons, whereas those who opt for breast feeding give reasons concerned with the welfare of the child. Another important factor is the response of the mother to breast feeding. The physiological responses to coitus and breast feeding are closely allied, and the act of breast feeding may give rise to embarrassment and feelings of guilt. Indeed, studies have shown that mothers who bottle feed tend to be more sexually inhibited and show signs of 'psycho-sexual disturbances' more frequently than those who breast feed. Women experiencing difficulties in labour also tend to have breast-feeding problems. The child itself may also be a cause of lactation failure if it suckles inefficiently, either because it is suffering from the effects of barbiturate sedation given to the mother at labour or because it is also being taught the technique of sucking milk from a bottle which is not appropriate to obtaining milk from the breast. In the category of group-derived emotions are included the effects of education and social class – breast feeding being more common at the higher levels of both. Important also is the attitude of the husband, of members of the family and of the family doctor. All these factors determine success or failure. There is evidence that many of these socio-psychological factors may depress lactation by upsetting the milk-ejection reflex. However, in view of the influence certain regions of the cerebral cortex may exert on prolactin secretion, as well as the effects of visual, auditory and olfactory stimuli, as revealed in animal studies, we

Lactation and its hormonal control

may infer that many of the above complex psychological factors also depress lactation by influencing the release of anterior pituitary hormones. Clearly more research is required both on the effects of socio-economic conditions on the breast-feeding pattern of different human population groups and the psycho-physiological mechanisms that link the emotion of the mother with milk secretion and milk ejection.

I end this chapter with further thoughts from a WHO report on the physiology of lactation: 'From the biological point of view, it is interesting to speculate whether deprivation of breast feeding has the same effects as deprivation of maternal care and whether the removal of the species-survival value of the capacity for lactation will in time lead to the retention of the human breast only as a sex symbol.'

SUGGESTED FURTHER READING

The Exploitation of the Milk-Ejection Reflex by Primitive Peoples. E. C. Amoroso and P. A. Jewell. Occasional paper no. 18 of the Royal Anthropological Institute, 126 (1963).
Physiology of Lactation. A. T. Cowie and J. S. Tindal. Physiological Society Monograph. London; Edward Arnold (1971).
Feeding the newborn baby. P. A. Davies. *Proceedings of the Nutrition Society* **28,** 66 (1969).
The milk-ejection reflex: a neuroendocrine theme in biology, myth and art. S. J. Folley. *Journal of Endocrinology* **44,** x (1969).
Organ culture techniques and the study of hormone effects on the mammary gland. I. A. Forsyth. *Journal of Dairy Research* **38,** 419 (1971).
Milk-ejection activity (oxytocin) in peripheral venous blood during lactation and in association with coitus. C. A. Fox and G. S. Knaggs. *Journal of Endocrinology* **45,** 145 (1969).
Echidnas. M. Griffiths. International Series of Monographs in Pure and Applied Biology, Zoology Division, vol. 38. Oxford; Pergamon Press (1968).
Psychologic aspects of lactation. N. Newton and M. Newton. *New England Journal of Medicine* **277,** 1179 (1967).
Morphogenesis of the mammary gland. A. Raynaud. In *Milk: the Mammary Gland and its Secretion*, vol. 1, chap. 1. Ed. S. K. Kon and A. T. Cowie. New York and London; Academic Press (1961).

Suggested further reading

Foetal development of the mammary gland and hormonal effects on its morphogenesis. A. Raynaud. In *Lactation*. p. 3. Ed. I. R. Falconer. London; Butterworths (1971).

Reproductive physiology of marsupials. G. B. Sharman. *Science, New York* **167**, 1221 (1970).

Physiology of Lactation. World Health Organization. World Health Organization Technical Report Series, no. 305. W.H.O. Geneva (1965).

Index

Index